Lecture Notes in Artificial Intelligence 6505

Edited by R. Goebel, J. Siekmann, and W. Wahlster

Subseries of Lecture Notes in Computer Science

FoLLI Publications on Logic, Language and Information

Alain Lecomte   Samuel Tronçon (Eds.)

Ludics, Dialogue and Interaction

PRELUDE Project – 2006-2009
Revised Selected Papers

Springer

Volume Editors

Alain Lecomte
Université Paris 8, Laboratoire Structures Formelles du Langage
Bat D salle 324, 2, Rue de la Liberté, 93200 Saint-Denis, France
E-mail: alain.lecomte@univ-paris8.fr

Samuel Tronçon
Résurgences/The NetLab.org
111, Rue Consolat, 13001 Marseille, France
E-mail: stroncon@resurgences.eu

ISSN 0302-9743 e-ISSN 1611-3349

ISBN 978-3-642-19210-4 e-ISBN 978-3-642-19211-1

DOI 10.1007/978-3-642-19211-1
Springer Heidelberg Dordrecht London New York

Library of Congress Control Number: 2011920711

CR Subject Classification (1998): I.2, H.4, F.4.1, F.1, F.3, G

LNCS Sublibrary: SL 7 – Artificial Intelligence

Typesetting: Camera-ready by author, data conversion by Scientific Publishing Services, Chennai, India

Printed on acid-free paper

Springer is part of Springer Science+Business Media (www.springer.com)

Preface
The PRELUDE Program

The articles collected in this volume are based on contributions to workshops and meetings that were held within the context of the PRELUDE project. PRELUDE, an acronym for "Towards Theoretical Pragmatics based on Ludics and Continuation Theory," ran from November 2006 to November 2009, with funding from the new French National Agency for Research (ANR). The objective of the project was to develop perspectives on natural language semantics and pragmatics based on recent developments in logic and theoretical computer science.

In the history of logic, research on the conditions for a theoretical discourse to be coherent and rigourous, mainly initiated by G. Frege, was followed in the mid-twentieth century by a new approach, inspired by the Computing Revolution, focusing on the study of programs and their transformations. The view according to which programs could be seen as the proofs of the statements expressing their correctness made logical formalisms into valuable objects of study for computer science. Among such logical frameworks, intuitionistic logic was considered as an attractive system because of its constructiveness, which takes the form of a famous correspondence between programs and proofs, known as the Curry–Howard correspondence (Howard 1980). After this discovery, the possibilities of extending such a correspondence were explored, and denotational semantics began to give more attention to the interpretation of proofs than to the interpretation of formulae.

Around the beginning of the 1990s, it became clear, mainly through the seminal work of A. Ranta (Ranta 1994) based on Martin–Löf's type theory (Martin-Löf 1984), that the semantics of proofs could provide the foundations for a view on natural language semantics going beyond the simple truth-conditional approach usually taken within the Montagovian tradition. Constructive type theory, for instance, turned out to be a suitable basis for an account of anaphora in so-called *donkey sentences*, such as *every farmer who owns a donkey beats it*. Nevertheless, although the view of a sentence meaning as the set of its proofs allows for more fine-grained distinctions than a strict Fregean conception, it remains a *static* view. *Interaction*, which is the key point of *dialogue*, was still ignored. Pragmatics in the type theoretical frameworks remains concentrated on the agent's judgements, which are viewed as mainly *static*.

The frameworks of linear logic (Girard 1987; Curien 2004), and then ludics (Girard 2001; Girard 2003, Girard 2006; Curien 2004), developed by Jean-Yves Girard, have provided new ways of thinking about dynamic processes and interaction. The objective of the PRELUDE project was to take these into account in order to analyze argumentation, dialogue, semantics and speech acts.

- M. Basaldella, A. Saurin and K. Terui bring innovative tools (c-designs) to extend ludics into a real framework for rebuilding the concepts of computability.

Other papers are devoted to topics which surround ludics, in that they provide:
- A perspective to situate ludics with regards to game theoretical frameworks, via a philosophical reflection on the content of the logic games (T. Tulenheimo)
- Concrete illustrations of the utility of tools to represent interactive situations as they occur in the dynamics of language (R. Kempson, Gregoromichelaki and Meyer-Viol), or the process of discourse (G. Winterstein and G. Schaden)
- Alternative frames to represent commitments (F. Cardone)
- Other alternative frames to represent interaction in the case of questions and answers (J. Groenendijk and F. Roelofsen)

Our first objective is actually reflected in this only by two papers, one of which is rather critical about the use of continuations even if it seems that the position it adopts with regards to questions like scope ambiguities is equivalent to a solution using them (C. Pollard). The other one (M. Petrolo) is more oriented toward theoretical computer science and gives a survey of various views of *duality*. Finally, another topic close to ludics is discussed, concerning *coherence spaces* (used as a semantics for linear logic), and their possible applications to ontologies (M. Romano, C. Fouqueré and M. Abrusci).

The workshops were held in Carry-le-Rouet (June 2007), Pauillac (October 2007), Autrans (May 2008), Hamburg, as part of ESSLLI, (August 2008), and the final colloquium was held in September 2009 in Paris. During these meetings several major scientists contributed, including (among researchers not directly included in the program): M. Abrusci, N. Asher, C. Beyssade, J-L. Dessalles, C. Faggian, J-Y. Girard, H. Herbelin, J-B. Joinet, L. Keiff, R. Kempson, M. Marion, P-A. Melies, D. Miller, M. Moortgat, A. Pietarinen, C. Pollard, A. Ranta, G. Sandu, L. Strassburger, T. Tulenheimo, G. White. The PRELUDE project was coordinated by Alain Lecomte. Institutions involved were the Mixed Research Units (UMR) "Formal Structure of Language" (University Paris 8), "Institute of Mathematics of Luminy (Aix-Marseille 2), LORIA (Nancy) and LABRI (Bordeaux) and, in the last period, the UMR LIPN (Paris-Nord). The project began in November 2006 and lasted 36 months. It benefited from an ANR funding of 147, 397 euros and hired a postdoctoral researcher for for 2 years (Samuel Tronçon).

The editors of this volume hereby thank all the members of these teams for their enthusiastic collaboration, and particularly Marie-Renée Fleury and Myriam Quatrini (IML), Laurent Roussarie (SFL), Christophe Fouqueré (LIPN), Christian Retoré, Richard Moot, Sylvain Salvati (Labri), Philippe de Groote, Sylvain Pogodalla (LORIA).

April 2010 Alain Lecomte

References

[Andréoli 1992] Andréoli, J.-M.: Logic Programming with Focusing Proofs in Linear Logic. The Journal of Logic and Computation 2(3), 297–347 (1992)

[Asher & Lascarides 1998] Asher, N., Lascarides, A.: Questions in dialogue. Linguistics and Philosophy 21, 237–309 (1998)

[Barker 2002] Barker, C.: Continuations and the nature of quantification. Natural Language Semantics 10, 211–242 (2002)

[Barker 2004] Barker, C.: Continuations in Natural Language. In: Thielieke, H. (ed.) Proceedings of the Fourth ACM SIGPLAN Continuations Workshop, pp. 1–11 (2004)

[Bernardi & Moortgat 2007] Bernardi, R., Moortgat, M.: Continuation semantics for symmetric categorial grammar. In: Leivant, D., de Queiroz, R. (eds.) WoLLIC 2007. LNCS, vol. 4576, pp. 53–71. Springer, Heidelberg (2007)

[Curien 2004] Curien, P.-L.: Introduction to linear logic and ludics, part I and II (to appear), downloadable from `http://www.pps.jussieu.fr/~curien/LL-ludintroI.pdf`

[Curien & Herbelin 2000] Curien, P., Herbelin, H.: Duality of computation. In: Proceedings of the Fifth AGM SIGPLAN, Montreal (2000)

[Danos, Joinet & Schellinx 2003] Danos, V., Joinet, J.-B., Schellinx, H.: Computational isomorphisms in classical logic. Theoretical Computer Science 294, 353–378 (2003)

[Faggian 2006] Faggian, C.: Ludics and interactive observability: the geometry of tests. Theoretical Computer Science 350, 213–233 (2006)

[Fleury-Quatrini 2004] Fleury, M.-R., Quatrini, M.: First order in Ludics. Mathematical Structures in Computer Science 14(2), 189–213 (2004)

[Fouqueré 2009] Fouqueré, C.: Ludics and Web: Another Reading of Standard Operations. Electronic Notes in Theoretical Computer Science, `http://www.elsevier.nl/locate/entcs`

[Girard 1987] Girard, J.-Y.: Linear Logic. Theoretical Computer Science 50 (1987)

[Girard 1999] Girard, J.-Y.: On the Meaning of Logical Rules-I. In: Berger, U., Schwichtenberg, H. (eds.) Computational Logic. Springer, Heidelberg (1999)

[Girard 2001] Girard, J.-Y.: Locus Solum. Mathematical Structures in Computer Science 11, 301–506 (2001)

[Girard 2003] Girard, J.-Y.: From Foundations to Ludics. Bulletin of Symbolic Logic 09, 131–168 (2003)

[Girard 2006] Girard, J.-Y.: Le Point Aveugle, vol. I, II. Hermann, Paris (2006)

[de Groote 2001] de Groote, P.: Type raising, continuations and classical logic. In: van Rooy, R., Stokhof, M. (eds.) Proceedings of the Thirteenth Amsterdam Colloquium, pp. 97–101 (2001)

[de Groote 2006] de Groote, P.: Towards a Montagovian Account of Dynamics. In: Proceedings of Semantics and Linguistic Theory, SALT XVI, Tokyo (2006)

[Gunlogson 2003] Gunlogson, C.: True to Form: Raising and Falling Declaratives as Questions in English. Routledge, New York (2003)

[Heim 1988] The semantics of definite and indefinite noun phrases. Garland Pub., New-York (1988)

[Hamblin 1970] Hamblin, C.-L.: Fallacies. Vale Press, Newport News (republished in 2004)

[Herbelin 2005] Herbelin, H.: Au coeur de la dualité, PhD thesis, University of Paris XI (2005)

[Hintikka-Kulas 1983] Hintikka, J., Kulas, J.: The Game of Language: Studies in Game Theoretical Semantics and its Applications. D. Reidel, Dordrecht (1983)

[Hintikka-Sandu 1997] Hintikka, J., Sandu, G.: Game Theoretical Semantics. In: Van Benthem, J., ter Meulen, A. (eds.) Handbook of Logic and Language, ch. 6. Elsevier, Amsterdam (1997)

[Howard 1980] Howard, W.A.: The Formulae-as-Types Notion of Construction. In: Seldin, J.P., Hindley, J.R. (eds.) To H. B. Curry. Essays on Combinatory Logic, Lambda Calculus and Formalism, pp. 479–490. Academic press, London (1980)

[Kamp 1981] Kamp, H.: A Theory of Truth and Semantic Representations. In: Groenendijk, J., Janssen, T., Stokhof, M. (eds.) Formal Methods in the Study of Language, Mathematical Centre Tract 135, Amsterdam, pp. 277–322 (1981)

[Kempson, Meyer-Viol & Gabbay 2001] Kempson, R., Meyer-Viol, W., Gabbay, D.: Dynamic Syntax: The Flow of language Understanding. Blackwell, Malden (2001)

[Lorenzen 1960] Lorenzen, P.: Logik und Agon, in Atti del XII Congresso Internazionale di Filosofia, Venezia (1958); reprinted in Lorenzen and Lorenz Dialogische Logik, Wissenschaftliche Buchgesellschaft (1978)

[Martin-Löf 1984] Martin-Löf, P.: Intuitionistic Type Theory, Bibliopolis, Naples (1984)

[Parigot 1992] Parigot, M.: $\lambda\mu$-calculus: an algorithmic interpretation of classical natural deduction. In: Voronkov, A. (ed.) LPAR 1992. LNCS, vol. 624. Springer, Heidelberg (1992)

[Pietarinen 2007] Pietarinen, A.-V.: Game Theory and Linguistic Meaning. Elsevier, Amsterdam (2007)

[Pinker 2007] Pinker, S.: The Stuff of Thought, Language as a Window into Human Nature. Viking, New York (2007)

[Rahman & Keiff 2005] Rahman, S., Keiff, L.: On how to be a dialogician. A short overview on recent development on Dialogues and Games. In: Vanderveken, D. (ed.) Logic, Thought and Action, pp. 1–51. Springer, Heidelberg (2005)

[Ranta 1994] Ranta, A.: Type-Theoretical Grammar. Oxford University Press, Oxford (1994)

[Shan 2004] Shan, K.: Delimited continuations in Natural Language: quantification and polarity sensitivity. In: Thielieke, H. (ed.) Proceedings of the Fourth ACM SIGPLAN Continuations Workshop, pp. 1–11 (2004)

[Sundholm 1986] Sundholm, G.: Proof Theory and Meaning. In: Gabbay, D., Guenthner, F. (eds.) Handbook of Philosophical Logic, vol. III, pp. 471–506. D. Reidel, Dordrecht (1986)

[Tronçon 2006] Tronçon, S.: Dynamique des démonstrations et théorie de l'interaction, PhD thesis, Université d'Aix-Marseille (2006)

[Wadler 2003] Wadler, P.: Call-by-value is dual to call-by-name. In: Proceedings of the International Conference on Functional Programming, ICFP 2003, Uppsala (2003)

[Walton 2000] Walton, D.: The Place of dialogue theory in logic, computation, and communication studies. Synthese 123, 327–346 (2000)

[Wittgenstein 1953] Wittgenstein, L.: Philosophische Untersuchungen. Blackwell, Malden (1953)

Organization

This volume was organized by the PRELUDE program and directed by Alain Lecomte and Samuel Tronçon. This work was supported by the french National Agency for Research, project no. 06-BLAN-0032 "PRELUDE."

Lecture Committee

Speech Acts in Ludics*

Marie-Renée Fleury and Samuel Tronçon

Institut de Mathématiques de Luminy - Aix-Marseille Université

Abstract. In this paper, we attempt to show that recent developments in proof theory, especially with ludics, are relevant for the study and the formalization of speech acts. This logical framework does not deal with truth values but with proofs, and this opens a new way for taking in charge the performative part of linguistic utterances. After having presented two models of speech acts and what theoretical elements we will hold as relevant for our own model, we introduce the ludical point of view by defining a speech acting conceptualization which renders some determinations not presented in the former models. We end by giving some examples of speech acts, presented in their ludical embedding, and we discuss what features the model provides.

Our aim is to show that recent developments in proof theory, especially with ludics, are relevant for the study and the formalization of speech acts. A common view about proof theory is that this logical framework does not deal with truth values but with proofs, and this opens a new way for taking in charge the performative part of linguistic utterances. Proofs are sometimes treated in the litterature as a syntactic objet, and beotians ignore what could be a semantic for proof theory. According to the cut elimination procedure due to Gentzen, and the related properties of convergence (the so called "*Hauptsatz*"[1]) and confluence (Church-Rosser), "*denotational semantics*" are defined only upon proofs with proof reduction. And a recent and genious extension of these ideas is given by the notion of "*encounter*" in Ludics, due to Girard [13], by which we can study convergences and divergences in the process of interaction between dialogical structures[2].

We present in the first section two models of speech acts and what theoretical elements we will hold as relevant for our own model. Secondly, we introduce the ludical point of view by defining a speech acting conceptualization which renders some determinations not presented in the former models. In the third section we present the ludic framework, not extensively detailed but focusing on the main points used in our formalization (designs, behaviors and the normalization procedure). Last but not least, we give some examples of speech acts, presented in their ludical embedding, and we discuss what features the model provides.

* This work was supported by the ANR project no. 06-BLAN-0032 "PRELUDE".

[1] The "*Hauptsatz*" is a theorem proved by Gentzen, which ensure the convergence of calculus (by the process of cut-elimination). This convergence is strong in the case of an intuitionnistic calculus, and weak for the classical one.

[2] For a good introduction to actual development of proof theory and the works of Girard, you can read recent works of Jean-Baptiste Joinet [17] and Samuel Tronçon [28]..

Lecomte and S. Tronçon (Eds.): PRELUDE 2010, LNAI 6505, pp. 1–24, 2011.

1 Introducing Speech Acts

1.1 Classical View

First we observe that foundationnal remarks about speech acts and the classical separation between constative and performative assumptions, as given first by Austin and Searle, opens the way to a bridge with proof theory. For example, the meaning of a constative sentence like %`The weather is good` can be explained in terms of truth and falsity. But, in the case of a performative sentence like %`I wish you to grow old`, we can't say what would be the truth value. We just know that, in some cases depending of the context, the sentence would be understood as a real wish or not. For this reason, we can identify a whole class of natural sentences which are not truth valuable and for whose meaning determination require a contextual analysis : these utterances convey "*Speech Acts*", *i.e.* (linguistic) objects which realize some concrete (pragmatic) action. A short analysis of embodied speech acts shows that their realizability conditions are unhomogeneous. Some of them are very primitive (*for expl.* the presence of an addressee which can receipt messages), others are conventional (*for expl.* use of water for the baptism) and procedural (*for expl.* to get divorced is possible only after being married). We have also conditions justified by mental states (*for expl.* sincerity of the promiser), and behaviors or expectations accorded with actual speech act orientation (*for expl.* in the case of the promise, the adressee would expect for the promised object).

We could define the core problem of the speech acts theory as a problem of assignment of type (Gazdar, 1981) : what markers and processes do we use to recognize the category of a speech act ? In the tradition, a large place is given to illocutionary force markers (*for expl.* %`I promise ...` in a promise), because it is the most evident manner for typing a speech act. But, when considering actual speech acts, explicit and/or pure syntactic markers are not sufficient, and we would have to consider not explicit markers, pragmatic means and contexts to specify the type of a speech act.

In the most common view, we can define a speech act as a mean used by a locutor in order to produce an effect in its environment by its words. By this use, he wants to inform, incite, convince or demand something to its addressee.

We suppose that for each meaning M, there exist an expression E which is a relevant formulation of M, *i.e.* E is sufficient to transmit the meaning's intention M to the interlocutor. So, following the *founding fathers*[3], we could define a speech act as a quadruplet (I,Γ,B,Υ) given by a communicational intention (I), a set of pre-requisite conditions (Γ), a body which realize concretely the action (B), a set of effects (Υ). This definition states speech acts in their purest form, *i.e.* the most disembodied one, as we see in the following example.

Example 1 (The promise). A speech act S will be considered of the type "promise" when four conditions are respected[4] :

[3] Evidently, Austin, Searle and their successors.

[4] Considering that the object of this article is not to introduce precise categorizations of speech acts but a reflexion on their formalization, we will not consider in our examples all the possible conditions, just their most common specificities.

- *Spk* has the intention of meaning some propositional content *p*
- *Spk* wants to make a commitment to *Add* about *p*
- *Spk* really wants to fulfil this commitment
- for *Spk* and *Add* , it is an evidence that without this speech act, *Spk* would not have done the promised action

Some remarks can be done at this stage. First, we do not know exactly how to consider and take in charge the speaker's **intention** of communicate something because the speech act is defined as if the intention was totally transparent.

Second, the **normativity** of speech acts is unclear, because it suggests that conditions are fully established before the realisation of the speech act. But we know that by convention we intend to justify the fact that speakers associate linguistics expressions with meanings they want to refers.

Last, the definition suffers from a lack of clarity and precision. It concerns notably the fact that we do not know exactly where a speech act begins : Is the speech act distinct from its effects ? What difference can we make between the effect and the intention ?

Considering these critical points, we consider that a good theory of speech act would have to define conventions and intentions not only as core notions but, evidently, as material objects upon which interactions between agents can be developed. Although these critical points, which we will take in charge in the last part of this paper, we find here the main connections with our proof-view in logics. The body of a speech act, *i.e.* the means which operates an effect, can be seen as a function responding to an expectation. This function, in order to be operational (*i.e.* to produce some effect), must interact with another function, some set of data and/or contextual elements : we would call all this bunch of unhomogeneous elements *functionnal environment.*

Speech Acts Theory	**Functions Theory**
intention	function's type
conditions	contextual datas required in a given type
body	function
effects	output datas produced in a given type

1.2 Inside Games

One hypothesis about speech acts that we do not want to justify is about the *Literal Force Hypothesis* [18]. It establishes a bi-univoque relation between a restricted set of illocutionary forces and a closed set of clause types. We agree with, for example, Beyssade and Marandin [4] which wants to solve some problems of assignement in speech acts without assumption about litteral force [9]. For Beyssade and Marandin, we must enrich the semantics and take in charge in a more fine manner the process of communication. In other words, speech acts must be seen as actions committed (by a speaker) and accepted (by some addressee), which updates some databases about facts, things to do and commitments. This mirroring effect can be called the *game-play*, and the syncing effect commitments by the way of functions modifying some data bases. And by the use of ludics we introduce also a parallel analysis of speech acts, in which speech acting structures (contexts and executables) are homogeneously considered, as well identified by their polarity. In fact, we take the interaction at its more primitive level, like a sort of "machine language" which "talks" both the language of executables structures and the contextual structures one.

For another example, the inscriptions, when considered in the game view, are notifications in a notebook managed by some interacting agent (a gameboard). It suggests that commitments are countable, and that for each commitment there is almost two inscriptions in two personal notebooks, or just one inscription in some big shared notebook. This hypothesis holds very well, but in the case of a negative condition, as in the promise *for expl.* , when we have to verify that the speaker is not already commited to do the promised thing, we do not know exactly what could be an inscription of the fact that "*someone is not committed*"... A possible approach would define it as the fact that there is no inscription about that in the available notebooks. Agents would have to scan notebooks, searching for an inscription, exiting with an error. In a *ludical view* we consider that, for a good approximation of the problem, we can define the negative condition as the absence of opposition of the present speakers to the presupossed fact contained in the speech act. As if the speaker was saying "*I'm not actually commited to do the thing I want to realize with this promise*" and nobody would have neither arguments nor the desire to refute this proposition. The interest in this point is that "absence of reaction" is a procedural definition, because there could be no known reaction at t time, and a contra-reaction at $t+1$: speech act evaluation became a real-time conceptualization and the model is self-contained. So, the basic scheme we assume about contexts is summed up by the following table :

	Game view	*Ludical view*
negative condition	absence of inscription	latence of reaction
positive condition	presence of inscription	actance of reaction

Latence means that the speaker knows that some reactions are possible, but there is no reaction in this branch of the design structure, or there is an explicit giving up. *Actance* means that some reactions are activated on a branch of the design structure. Evidently, we must explain how the agent can know something by himself. We define the process of "knowing by itself" exactly as the process of "knowing by somebody", but directed on itself. So, knowing that a condition is realized from my own view is reduced to the fact of testing possible reactions of myself about some proposition.

2 Speech Acts in Ludics

As we must justify the relevance of a new model about speech acts, we present here the most important points for understanding our view, in their most intuitive form. The further sections being devoted to present the technical framework and some examples.

Before going into details of our conceptualisation of speech acts let us introduce some core notions, which are imported form ludics, and would be precisely defined in the next section.

2.1 Speech Acting

While speech acts are viewed as stable structures, we introduce a "speech acting" description in which the interaction plays the main part.

Classical speech acts are presented in a synchronous view. But objectively, the procedurality of embodied speech acts oblige us to introduce some complexity. Speech acting is the fact of an asynchronous processing, and we will take in account this fact first in our conceptualization, secondly in our formal view.

First, we recount the interaction process as a parallel processing, realized by two speaking agents, forming their own actions on the basis of their interlocutor's actions.

Process 1 (Speaker's action)

⋆ *Spk proposes a speech act* $\mathfrak{A}$. *It can be seen as a function defined on the context with values in a subset of the context containing the commitments of the interlocutors. The context must satisfy some conditions* $C_1, \cdots, C_n$ *in* C. *So we denote by* $\mathbb{C}$ *a closed*[6] *set filled with justifications of* C *(*$\mathbb{C}$ *will be called a behavior, i.e. a set of designs). As we just define* A, *we can write* $\mathfrak{A} \in \mathbb{C} \rightarrow \mathbb{E}$ *which means "there is a construction called* $\mathfrak{A}$ *which transforms the contextual elements in* $\mathbb{C}$ *to obtain contextual elements in* $\mathbb{E}$*".*

⋆ *Spk thinks that he can access in the (open) context to some locus in which the condition* C *is justified by a design* $\mathfrak{D}$ *: knowing that, he is considering a set* $\mathbb{C}$ *of justifications of* C. *This design can be located in a private or a (preferably) shared part of the context.*

⋆ *Spk feels himself justified to provoke the commitment* $\mathfrak{E}$ *which would be realized in the context* $\mathbb{E}$.

Process 2 (Addressee's reaction)

⋆ *Add perceives the speech act*

⋆ *Add does not have objectively the same view on the context that Spk had. For example, he can perceive differently the same situation, i.e. having more elements in* $\mathbb{C}$, *or even possessing arguments which contradicts a condition* $\mathbb{C}_i$. *He can also refuse arbitrarily the effectivity of a commitment even if the conditions were filled (for expl. in the order, Add can ignore the authority of Spk on him).*

⋆ *Add responds by forecasting the speaker's actions or by acting in some locus, or by introducing some elements which were not taken in charge by the design of Spk .*

The complexity of speech acting representation is dependent of two important problems :

[6] The notion of closure will then be clarified in the next section.

Structures we present get the form of a plan of actions and reactions. In a schematic view, actions are done by the speaker, reactions are done by the addressee. But, this pattern masks the fact that these structures renders the subtle distinction between what is planned, what is added/modified in the interaction process, and what is finally realized. In our formalization, we present speech acts as realized, *modulo* further continuations which left unchanged the given basic structure. So we must take in charge the difference between the action plan made by some speaking agent (the potent), parts of the action plan which are invoked at this time in the speaking situation (the patent), and open branchs which are not constructed as well at this time (the latent).

	of *Spk* for *Spk*	**of *Spk* for *Add***
latent	planned but not invoked	not explored but anticipated
potent	opened but not planned	not anticipated
patent	invoked	explored

We see speech acting as a parallel process implying two agent activities : the observation ("*what parts of the context are selected by my speech act's conditions of realization*") and the action ("*how I want to modify the present context*").The observation corresponds to what is anticipated by a locutor from its interlocutor (parts of its plan which have an inverse polarity). And the action corresponds to what is forecasted by some locutor from its own actions (parts of its plan which have the same polarity). The situation is slightly different depending on whether we are in a perfect world or a real world :

	perfect world *all is transparent*	**real world** *there is some opacity*
Spk observation	*Add* plan	negative steps of *Spk* plan
Spk action	*Spk* plan	positive steps of *Spk* plan
Add observation	*Spk* plan	positive steps of *Add* plan
Add action	*Add* plan	negative steps of *Add* plan

2.2 Speech Acts

Definition 1 (Ludical speech acts). *A "ludical speech act" is a sequence of reduction by which some executable (called the speech act) produce an impact when some conditions in the context are fulfilled.*

A type of "ludical speech act" is a class of executables which produces the same impact when observed in the same conditions.

The speech act is uttered in some determinated context which can contains facts, knowledges, past actions, mental states... Each speaking agent selects the parts of the context he considers as relevant, and acts by reference to these elements. For example, at the utterance of the speech act by *Spk* , *Add* will react by means of parts of the context he "perceives", and which could be different from those viewed by *Spk* . The difference between what views *Spk* and what uses *Add* plays a huge function in the realization of impacts, as *Spk* can feel himself so justified by contextual elements to obtain some effects which are refused by *Add* by means of some others contextual elements.

In the interaction, these contextual elements takes the form of possible actions of speaking agents in their plans. They are positive when invoked by *Add* as contexts (negatives for *Spk*), or invoked by *Spk* as conditions. They are negative when used by *Add* as an action (positives for *Spk*), or invoked by *Spk* as anticipation. Evidently, we take the classical conditions (preliminary and essential rules, commitments...) associated with the running speech act type as the main contextual elements invoked. In the case of the order, one of the preliminary conditions is that *Spk* had some authority on the actions done by *Add* , when he is for example his guru, his boss, his baby-sitter or his mother. Speaking agents refers to these elements essentially as argues in the dialogue, it is under this form that we will formalise them. For example, the set of actions associated with this first rule for the order contains the following utterances, which could be used in a real situation of speech acting dialogue :

- for *Spk* : %`I have authority on` *`Add`* , %`I'm the boss of` *`Add`* , %*`Add`* `is the child whom I am the babysitter`, %*`Add`* `is one of my children`...
- for *Add* (respectively) : %`I choosen to follow the lead of this man`, %`I'm not working`, %`She is not my mother !`, %`ok mum !`

Definition 2 (Structure of ludical speech acts). *A ludical speech act is defined by three elements :*

- *the* **speech act** *: some competence of the Spk for impacting the context considering some actual conditions or anticipated reactions of its interlocutor. This competence is invoked in the situation as a structured system of actions.*
- *the* **test** *: the interactive situation by which we oppose the speech act and a complex structure which plays the role of context mixing contextual datas with interlocutor reactions*
- *the* **impact** *: which is the result of the interaction, i.e. a modification of the context (inscription, erasing)*

So, we decompose the speech act in three levels : the body of the speech act (i.e. the speech act as we find in the litterature), the speech act in interactive situation (i.e. taking in account the reaction of addressee) and the achieved act (main or peripheral effect). In the following design we assume that the active part take the form of a function modifying context to produce an effect, which is the basic form underlying our models[7].

[7] Please consider that it is not exactly the case for all the speech acts situations we will present in the last section, by the fact that it depends on the type of speech act and the *modus operandi* of the interaction. But, as we say in french mathematics, it is "*morally the same*".

The designs are built by the means of only three rules schemes: two first schemes of rules are issued from logical rules (a positive one and a negative one)[9] and a new one called the damon ($\mathfrak{Dai}$), seen as a "Giving up". This rule does not arise from the logic, but is needed for taking in account the interaction.

Positive action : *to perform an action, to ask, to answer*
By a positive action, the player selects a branching locus and opens all its possible loci for continuing the interaction. He chooses to act in the place ξ and he opens to his interlocutor the range of actions $\xi * i_1 \vdash, \ldots, \xi * i_n \vdash$.

$$\frac{\xi * i_1 \vdash \Lambda_i \qquad \ldots \qquad \xi * i_n \vdash \Lambda_n}{\vdash \Lambda, \xi} (\xi, \{i_1, ..., i_n\})$$

Where $\cup \Lambda_i \subset \Lambda$.

REMARK: By playing an empty ramification, Player prevents Opponent from any reaction and so he blocks the continuation of the interaction.

Negative action : *to receive opponent action, to foresee*
Formally, a negative action specifies the ramifications of a directory. In our context, we can say that, in the place ξ, else after one of my action I get ready for receiving reactions of my interlocutor in a list that I have foreseen, or I receive a list of messages sent by my interlocutor.

$$\frac{\vdash \Lambda, \xi * i_1^1, \ldots, \xi * i_n^1 \qquad \ldots \qquad \vdash \Lambda, \xi * i_1^j, \ldots, \xi * j_m^j}{\xi \vdash \Lambda} \xi, \{R_1, ..., R_j\}$$

Where $R_k = \{i_1^k, \cdots, i_n^k\}$

Giving up : At any time in a positive context, the active player (the one who has to play a move) can prefer to give up ; so the immediate effect is to stop the interaction.

$$\frac{}{\vdash \Lambda} \dagger$$

Examples 1 ($\mathfrak{Dai}$ and $\mathfrak{Fax}$).
- $\mathfrak{Dai}^+$ and $\mathfrak{Dai}^-$ *I decide to immediately stop, as soon as I have the hand.*

$$\mathfrak{Dai}^+ = \frac{}{\Delta} \dagger \qquad \mathfrak{Dai}^- = \frac{\overline{\vdash \xi * I, \Delta} \dagger}{\xi \vdash \Delta} (\xi, \mathcal{P}_f(\mathbb{N}))$$

- The design $\mathfrak{Fax}_{\xi,\xi'}$ will play a crucial role in the following sections. As it is suggested by its name, this design is a sort of echolalic design. $\mathfrak{Fax}_{\xi,\xi'}$ allows us by the means of interaction to move a design $\mathfrak{D}$ localised at the address (locus) ξ into a design $\mathfrak{D}$' localised at the address ξ' ; it is recursevely defined. Rougthly speaking, it imitates the two first actions of the design $\mathfrak{D}$ and so on infinitely.

[9] The underlying logic is the hypersequentialized linear logic which works with polarized synthetic connectives. So only two rules schemes are needed.

Here $\mathcal{P}_f(\mathbb{N})$ is the set of finite subsets of $\mathbb{N}$; it suggests that all the possible cases are avalaible for the design to be delocalised.

$$\dfrac{\dots \quad \dfrac{\dfrac{\mathfrak{Fax}_{\xi' * i, \xi * i}}{\xi' * i \vdash \xi * i}}{\vdash \xi * I, \xi'}{\scriptstyle (\xi', I)} \quad \dots \quad \dfrac{\dfrac{\mathfrak{Fax}_{\xi' * j, \xi * j}}{\xi' * j \vdash \xi * j}}{\vdash \xi * J, \xi'}{\scriptstyle (\xi', J)} \quad \dots}{\xi \vdash \xi'}{\scriptstyle (\xi, \mathcal{P}_f(\mathbb{N}))}$$

From now $\underset{\xi \vdash \Delta}{\nabla}$ will symbolize the design (seen as a tree) based on $\xi \vdash \Delta$.

3.2 The Interaction

In Linear Logic, the interaction is represented by the cut elimination. In Ludics, the interaction is a meeting between two players strategies localised in a same locus[10]. The **dynamics of the interaction** is given by a process[11] which regulates this meeting. At any step, action-reaction of players are put in coincidence until one of the players gives up or until this coincidence failed or even endlessly continues.

Without entering the formalism, we can say that the interaction between designs is seen as a cut-net, *i.e.* an acyclic finite graph of designs pairwise connected by their bases. For example, let us consider the following cut-net of base[12] $\sigma \vdash \lambda, \rho$:

We don't give here a complete description of this procedure, as we just present two views about how it is processing :

Example 3. In this example the base of the cut-net[13] is $\vdash$. We can observe that in this example, the ramification $R = \{1, 3\}$ is in the directory $\mathcal{R} = \{\{1, 3\}, \{2, 3\}\}$

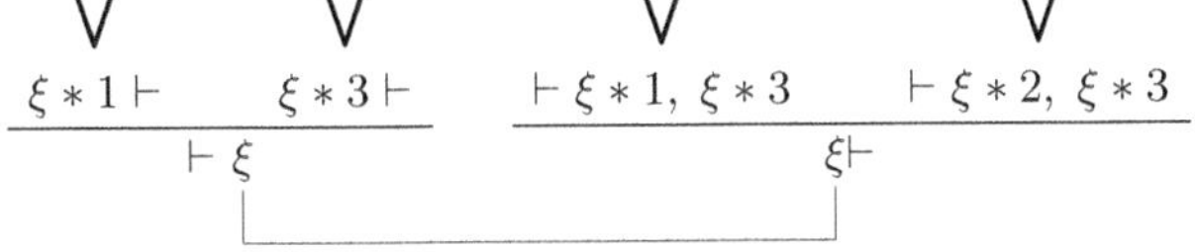

---

10 The interaction is concretely translated by a coincidence of two loci in dual positions in the bases of two designs. For example a design of base $\sigma \vdash \xi$ can interact against a design of base $\xi \vdash \rho$.

11 This process is called the "normalisation procedure" in lambda-calculus, or the "cut elimination procedure" in the sequent calculus, by logicians and computer scientists.

12 The base of the cut-net is obtained by erasing the cut loci, *i.e.* the loci by which they are connected.

13 such a cut-net is called a closed cut-net.

The first reduction step eliminates dead forks (the branches upon an unconnected node) and produce the following net:

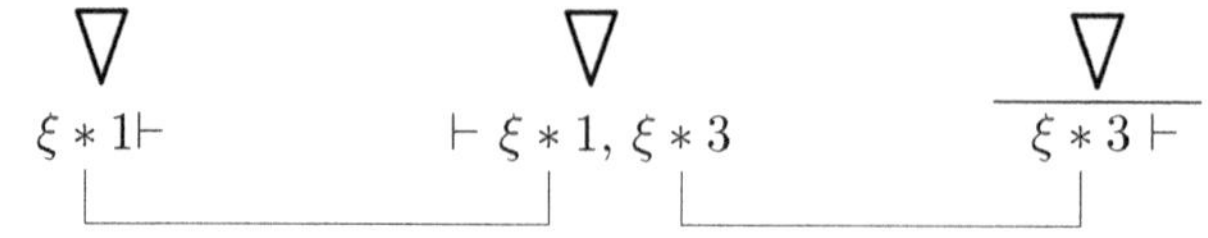

After n reduction steps, there's only three possible configurations :

- *Convergence* by "giving up": $\mathfrak{Dai}$ is one of the designs produced by the reduction, then all the net is reduced to the $\mathfrak{Dai}$.
- *Divergence* : the ramification R of the positive design is not in the directory $\mathcal{R}$ of the negative design.
- *Divergence* because of an infinite interaction : The exchange makes loops or continues infinitely.

Example 4. Here we present the interaction between a design $\mathfrak{D}$ of base $\vdash \xi$ and the $\mathfrak{Fax}$ of base $\xi \vdash \rho$. Let us observe that the base of this cut-net is not empty. In case of convergence, we then obtain a design $\mathfrak{D}'$ of base $\vdash \rho$; the resulting design in fact is similar to the design $\mathfrak{D}$.

To cut a design $\mathfrak{D}$ and the $\mathfrak{Fax}$ enables us to *delocalize* a design, to move it, to modify its place of anchoring.

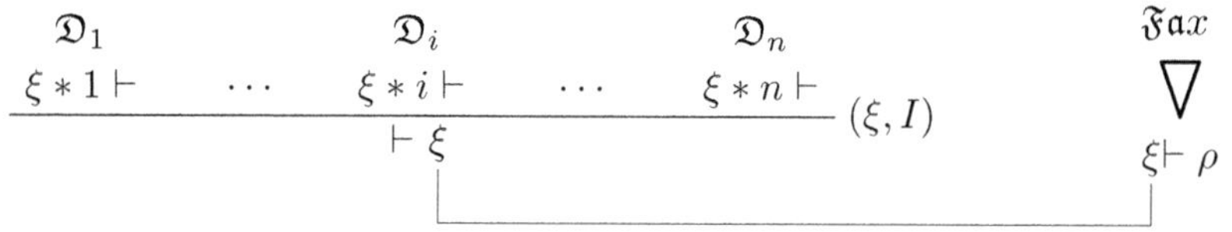

Two reduction steps produce the design:

$$\cfrac{\underset{\rho * 1 \vdash}{[[\mathfrak{Fax}_{\rho*1,\xi*1}, \mathfrak{D}_1]]} \quad \dots \quad \underset{\rho * i \vdash}{[[\mathfrak{Fax}_{\rho*i,\xi*i}, \mathfrak{D}_i]]} \quad \dots \quad \underset{\rho * n \vdash}{[[\mathfrak{Fax}_{\rho*n,\xi*n}, \mathfrak{D}_n]]}}{\vdash \rho}\ (\xi, I)$$

Finally, let us introduce the notion of orthogonality. It will allow the handling of complete objects named behaviors and presented in the further section. indexobjetOrthogonality.

Definition 1 (Orthogonality). *Two designs $\mathfrak{D}$ and $\mathfrak{E}$ are orthogonal if the reduction of the net $[[\ \mathfrak{D}\ ;\ \mathfrak{E}\ ]]$ terminates and produces $\mathfrak{Dai}$.*

3.3 Behaviors

Definition 2 (Behavior). *A behavior $\mathbb{C}$ is a set of designs of same base, closed by biorthogonal.*

$$\mathbb{C} = \mathbb{C}^{\perp\perp}$$

Then we can consider the following correspondence:

designs $\mathfrak{D}$	(para-)proofs π
behaviors $\mathbb{C}$	formulas A

At the abstract level, the behaviors can be composed by the means of the connectives of the LL:

$$\mathbb{C}_1 \otimes \mathbb{C}_2, \quad \mathbb{C}_1 \oplus \mathbb{C}_2, \quad \mathbb{C}_1 \multimap \mathbb{C}_2, \quad \mathbb{C}_1 \wp \mathbb{C}_2, \quad \mathbb{C}_1 \,\&\, \mathbb{C}_2, \quad \cdots$$

and then also be interpreted in interactive manner.

For example the behavior $\mathbb{C}_1 \multimap \mathbb{C}_2$ can be seen as a function which transforms a design in $\mathbb{C}_1$ into a design in $\mathbb{C}_2$. (remark: we also can speak of a design of type $\mathbb{C}_1$) We can also say that when a design $\mathfrak{E}_1 \in \mathbb{C}_1$ interact with a design $\mathfrak{D}$ in $\mathbb{C}_1 \multimap \mathbb{C}_2$, then the result is a design $\mathfrak{E}_2 \in \mathbb{C}_2$:

$$\begin{array}{ccccc} \mathfrak{E}_1 & \quad & \mathfrak{D} & & \mathfrak{E}_2 \\ \in & & \in & \leadsto & \in \\ \mathbb{C}_1 & & \mathbb{C}_1 \multimap \mathbb{C}_2 & & \mathbb{C}_2 \end{array}$$

Example 5. Let us consider a record in a data base with three fields : coordonnate (*cd*), shape (*sh*) and color (*col*). It can be represented by the following design $\mathfrak{R}$ based on $\xi \vdash$:

$$\dfrac{\dfrac{\dfrac{\dfrac{}{\vdash \xi * cd * 1 * 1}\emptyset}{\xi * cd * 1 \vdash}}{\vdash \xi * cd} \qquad \dfrac{\dfrac{\dfrac{}{\vdash \xi * sh * sqr * sqr}\emptyset}{\xi * sh * sqr \vdash}}{\vdash \xi * sh} \qquad \dfrac{\dfrac{\dfrac{}{\vdash \xi * col * yell * yell}\emptyset}{\xi * col * yell \vdash}}{\vdash \xi * col}}{\xi \vdash}$$

Suppose that you want to know the color (blue, red or yellow) written on the record. Let $\mathfrak{Q}$ be the design which represent the question: "which is your color ?". How does this design is built ? : first you ask the question, secondly you wait for three possible colors. Then, above any possibilities, you are getting ready to copy out the colour at the address $\sigma * col$. The corresponding design $\mathfrak{Q}$ is here :

$$\dfrac{\dfrac{\dfrac{\dfrac{\dfrac{}{\vdash \xi * col * blue, \sigma * col * blue}\dagger}{\sigma * col \vdash \xi * col * blue}}{\vdash \xi * col * blue, \sigma} \qquad \dfrac{\dfrac{\dfrac{}{\vdash \xi * col * red, \sigma * col * red}\dagger}{\sigma * col \vdash \xi * col * red}}{\vdash \xi * col * red, \sigma} \qquad \dfrac{\dfrac{\dfrac{}{\vdash \xi * col * yell, \sigma * col * yell}\dagger}{\sigma * col \vdash \xi * col * yell}}{\vdash \xi * col * yell, \sigma}}{\xi * col \vdash \sigma}}{\vdash \xi, \sigma}$$

Third case : As previously, Adrian and Sophie do not have the same knowledges, the context of Adrian is not identical to the one of Sophie.

Adrian does not want going to the movie tonight because he prefers to watch the World Cup final live on TV. So Sophie is mistaken. The interaction of Adrian and Sophie seems to fail; Sophie's promise fails, her commitment isn't recorded. Nevertheless a new knowledge is produced. It is put in the context shared by the mother and her son.

- Sophie : *"I promise you the movie tonight"*
- Adrian : *"I prefer watch the World Cup final live on TV!"*
- Sophie : *"Sorry, I have forgotten it."*

In this situation, the speaker is retracting. In Ludics, we represent this by allowing the speaker to "re-play" on a locus ξ. We say that we use "multi-addresses", as introduced by M. Basaldella and C. Faggian in [2]. We denote $\vec{\xi}$ a multi-address, it is in fact a sequence of addresses, ξ is the first element of it . All is working as if ξ was not consumed and remains in the sequence of available locus. In our case, when Sophie plays again on ξ, then she plays the $\mathfrak{Fax}$ to record this new information.

- First part of the play:

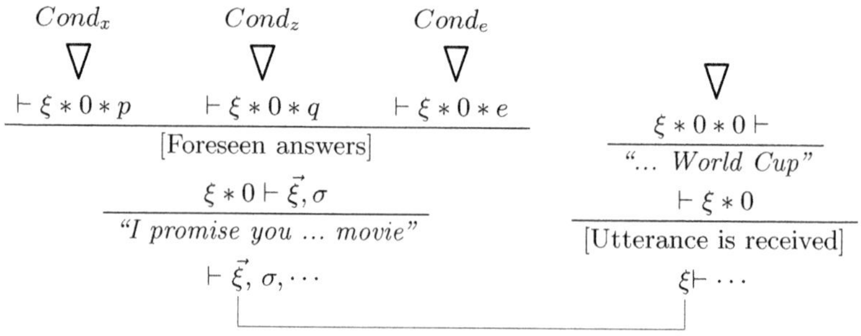

- Second part of the play:

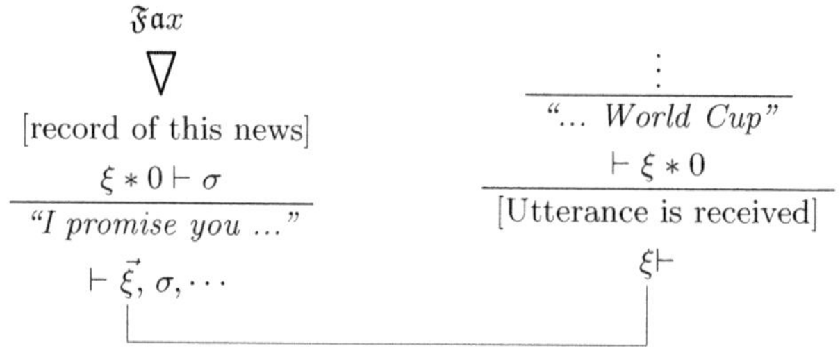

which is reduced in :

$$\frac{\maltese}{\vdash \sigma}\ \text{"... World Cup"}$$

4.2 Polite Request: "Can You Close the Window, Please"

This speech act has an interrogative grammatical form, but it must be understood as a polite request, perhaps as an order. Our aim is to show how the *interaction* between two agents can help us to determine if the act was well received as a request and not an interrogation; the interaction allows us to recognize the "felicity" of the act and then its categorization.

We will give the design associated with each of these interactive situations.

In order to simplify the example, we only consider the following conditions of felicity for a polite request (preliminary conditions and essential ones):

- $\mathbb{C}ond_P$: Sophie may ask Adrian performing the action (she observes the rules of politeness, she has authority over him, ...)
- $\mathbb{C}ond_M$: Adrian has the ability to perform the act
- $\mathbb{C}ond_R$: The act has not yet been achieved
- $\mathbb{C}ond_O$: After the statement, Adrian is committed to perform the act

Let us imagine several possible dialogues between Adrian and Sophie. They are tests allowing us to establish the happiness (or unhappiness) of the act.

First case : Where the felicity of the act is obtained. The normalisation succeed. Adrian is now commited to perform the act (to close the window). From the Ludic point of view, a design representing the commitment is created.

- Sophie: *"Can you close the window, please"*
- Adrian: *"OK, I do it immediatly"*
- Sophie: *"Thanks"*

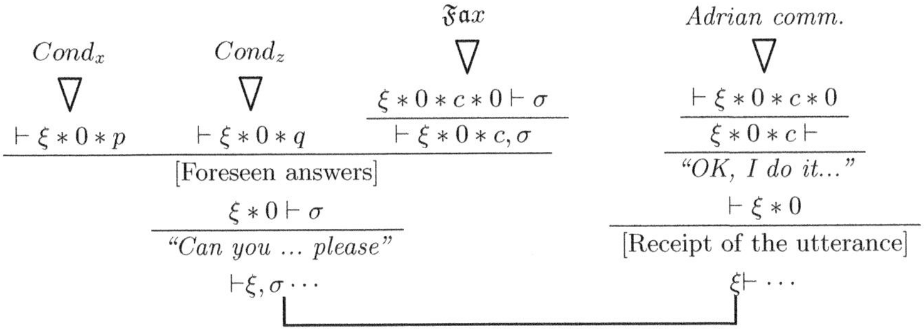

which is reduced in the following design we can take as the result of the speech act :

(copy of)

Adrian commitment

$$\nabla$$

$$\vdash \sigma$$

Second case : The felicity fails due to the insatisfiability of the condition $\mathbb{C}ond_A$ which say that "Adrian is able to perform the act". We could have the following interaction :

- Sophie: *"Can you close the window, please"*
- Adrian: *"I can't, I'm too little"*
- Sophie: *"Really? I did not know"*

Sophie don't wait for this answer from Adrian because she actually thought that Adrian was able to perform the act. She has justifications of the condition $Cond_A$, i.e. the design representing her strategy uses delocalisation of designs in the behavior $\mathbb{C}_A$. Consequently, the act can't be categorized as a polite request. As previously (in the third case of promise), Sophie has to replay on ξ to recognize her error and record this new knowledge.

- First step of the interaction:

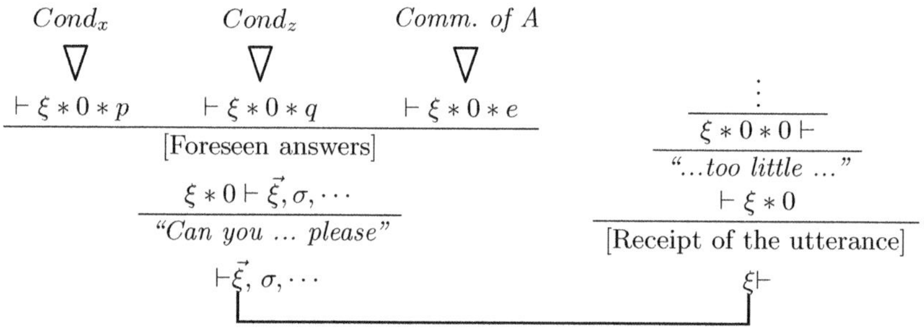

- Second step of the interaction:

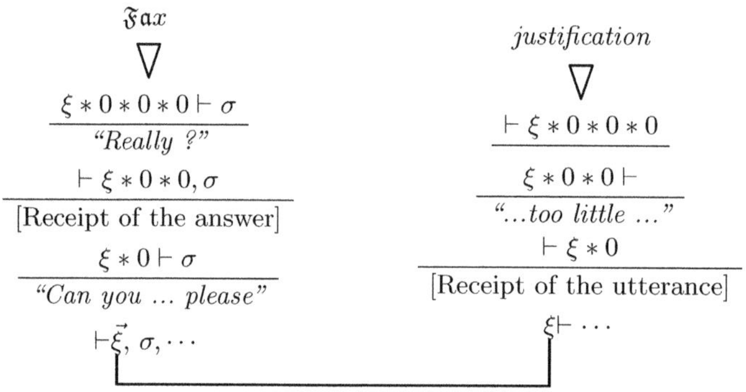

which is reduced in the following design :

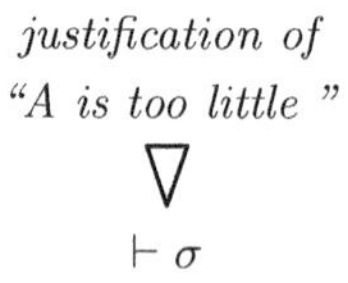

By playing on $\xi * 0 * 0$, Adrian stop the waited effect of the act which would be his commitment to perform the act (to close the window). We'll say that this speech act fails, it cannot be categorized as a polite request, due to the lack of ability of Adrian to perform the act.

The result of the interaction between the designs of Sophie and Adrian is the new knowledge of Sophie: *Adrian is more little than she thought.*

Third case : The unhappyness of the speech act is due to the insatisfiability of the condition $\mathbb{C}ond_R$: [The act isn't already performed].

- Sophie: *"Can you close the window, please"*
- Adrian: *"It is already closed"*
- Sophie: *"I don't understand, I've just opened it five minutes ago"*
- Adrian: *"Look at yourself"*
- Sophie: *"OK, I agree you"*

Adrian begins by attacking Sophie, because he sees that the condition $\mathbb{C}ond_R$ isn't realized i.e. he answers on a locus where Sophie has a justification of the fact that the window is open ("I opened it, only five minutes ago"). The dialogue could end by giving up of Adrian. He prefers to insist and so Sophie can know that the window is already closed.

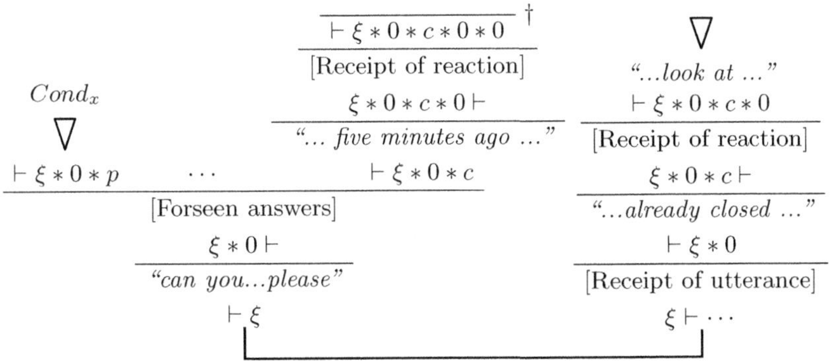

is reduced in $\mathfrak{Dai}$.

Fourth case : The unhappyness of the speech act is due to misunderstanding of the speech act: the *polite request* is understood as a *question.*

- Sophie: *"Can you close the window, please"*
- Adrian: *"I can't answer your question".*

Sophie did not ask a question, but makes a polite demand. Thus, we end in a failure (of the process of normalization), because Sophie did not foresee answer to be given to him.

16. Hamblin, C.L.: Mathematical Models of Dialogue. Theoria 37, 130–155 (1971)
17. Joinet, J.-B.: Logique et Interaction, Habilitation à diriger des recherches. Université Paris 7 (2007)
18. Sadock, J.M.: Toward a Linguistic Theory of Speech Acts. Academic Press, New-York (1974)
19. Searle, J.R.: Speech Acts. Cambridge University Press, Cambridge (1969)
20. Searle, J.R.: Indirect speech acts. In: Davis, S. (ed.) Pragmatics: A Reader, pp. 265–277. Oxford University Press, Oxford (1975-1991)
21. Searle, J.R., Vandervecken, D.: Foundations of Illocutionary Logic. Cambridge University Press, Cambridge (1985)
22. Searle, J.R., Vandervecken, D.: A taxonomy of illocutionary acts. In: Gunderson, K. (ed.) Language, Mind and Knowledge, Mineapolis University Press (1985)
23. Seligman, J., Moss, L.: Situation Theory. In: van Benthem, J., ter Meulen, A. (eds.) Handbook of Logic and Language, pp. 239–309 (1997)
24. Stalnaker, R.C.: Assertion. In: Cole, P. (ed.) Pragmatics, pp. 315–332
25. Stalnaker, R.C.: Inquiry. MIT Press, Cambridge (1987)
26. Stalnaker, R.C.: On the representation of context. Journal of Logic, Language and Information 7(1), 3–19 (1998)
27. Stalnaker, R.C.: Context and Content: Essays on Intentionality in Speech and Thought. Oxford University Press, Oxford (1999)
28. Trônçon, S.: Dynamique des démonstrations et théorie de l'interaction. Thèse de Doctorat, Université de Provence (2006)
29. Truckenbrodt, H.: Sentence Type Meanings. Ms available at (2004), `http://www2.sfs.nphil.uni-tuebingen.de/~hubert/Home/papers/`
30. Wittgenstein, L.: Remarks on the Foundations of Mathematics. The MIT Press, Cambridge (1983)

Speech Acts and Ludics: Reacting to Breakdowns of Interaction

Pierre Livet

Département de Philosophie - Aix-Marseille Université

Introduction

The first idea of this paper is that speech acts (orders, promises, excuses) are not simply ways of doing something by the very act of telling what we are doing, or ways of giving a supposed illocutionary strength to language, but ways of repairing breakdowns of interaction (excuses) or preventing anticipated breakdowns (orders, promises, declarations). The second one is that Girards ludics could be a useful tool to represent formally the different kinds of speech acts so conceived.

Ludics are a way to represent interactions that could lead to the construction of a logical proof. Possible proofs are confronted to their counter-proofs, so to speak. If the symmetry between a proofs and its counter-proof can be ensured, if "convergence" is the case, one of the two has to quit the interaction game, and the remainder is the winner proof. Of course, breakdowns of interaction are divergences, not convergences. Ludics can represent explicitly some categories of divergences. The idea is then to try to match each formal representation of divergence in ludics to a kind of breakdown in language interaction, and to express basic speech acts as ways to repair divergence and come back to convergence.

In order to show that this idea could work, we need first to present ludics in a way that enhances its capacity to express interaction, and lead to grasp its formal ways of expressing different kinds of divergences as ways of expressing breakdown of interaction. Then we will try to show that the main categories of speech acts seen as related to different breakdowns of interaction can be represented formally by these kinds of divergence.

1 A Presentation of Ludics Orientated towards the Expression of Speech Acts as Ways of Repairing Interaction Breakdowns

Ludics can be seen as well as a way of answering to the question: "is this formula proven?", or as a way of formalizing interactions between a proponent and an opponent. There is a family resemblance between ludics and dialogic, but ludics is more ambitious and complex. One player in the interaction, the proponent, chooses an action among a set of possible actions. The opponent offers a new set of actions accessible from this action, and the interaction goes on, alternating

Lecomte and S. Tronçon (Eds.): PRELUDE 2010, LNAI 6505, pp. 25–31, 2011.

choices and offers. Choices are said to have a positive polarity, offers a negative one.

Polarity can be also assigned to connectors as well as to quantifiers. For example, "Every" is negative, "Some" is positive, since "every" offers you every elements in a set, and "some" assumes that some choice has been done. The additive disjunction, "Plus", is positive, as if you choose to develop the formula A in the sentence "A plus B", this choice may not lead you to the same premises (when going upstream in the proof tree by eliminating connectors) than the choice of B. On the contrary, the additive conjunction, &, is negative, as you can develop ad libitum A and B. The tensor $\otimes$ (multiplicative conjunction) is positive, as we have not the choice of what formula to develop into sub-formula, we have to develop the two formulas at the same time and level of the proof tree, etc.. Negative can be said reversible, and positive reversible.

As the negative connectors let you free to make the operations of developing the proof tree in whatever order you want, it is possible to put together the negative operations, and to alternate them with the positive operations (the irreversible ones) each time those appear needed. Ludics imply that the reorganization of the formulas in accordance with such an alternation has been already done. The proof tree is so built as a level implying a positive operation follows a level implying a negative one, and conversely. There is no more need of rules devoted to the elimination or introduction of specific connectors, as what matters is whether they are positive making an irreversible choice – or negative – offering the possibility of several reversible choices. What remains needed is only the alternation of making a choice versus giving choices, and the notations (addresses or locus) that make possible to identify what sub-formulas are processed and also from what steps of analysis and from what bigger formulas they are coming from.

Two rules are sufficient for developing the proof tree, a positive and a negative one. The positive rule makes a choice among the formulas of the right of the turnstile, on the side of the consequences, and combines the address of the focused formula with each address of its premises, on the left of the turnstile. The addresses of the premises have to be available in a "ramification". For example:

$$\frac{011 \vdash \qquad 012 \vdash 02}{\vdash 01, 02}$$

in accordance with (rule +, focus 01, ramification $\{1,2\}$). The negative rule starts by focusing on a formula on the left side of the turnstile and offers as consequences at the upper level of the proof tree all the possible combinations of the address of the focussed formula with the different addresses of the different available ramifications (the set of which is called a repertory). For example:

$$\frac{\vdash 01, 02 \qquad \vdash 01, 03}{0 \vdash}$$

(rule −, focus 0, repertory $[\{1,2\}\{1,3\}]$). Showing that a formula is proved is giving inferential justifications for this formula. The positive rule chooses a consequence and gives the different premises that are needed for the justification of this consequence. For a given premise, the negative rule gives all the consequences already justified that are the inferential possible bases of the justification of this premise.

So far ludics seem only to be a new and more compact way of building proof trees, the highest level of which is usually made of axioms. But ludics dispense us to use axioms as well as it dispenses us to write formulas we need only their addresses. If there are no longer axioms, how could we know that a formula is proved? Girards idea is to use the same alternation between the right (positive) and the left (negative) side of the turnstile to get a proof by the confrontation of a formula with what could be called its "counter-formula", its symmetrical image with respect to the turnstile. The two symmetrical formulas lead to two developments using the positive and negative rules- that are called the positive and the negative "designs". Between the addresses of the two developments or designs, cutting addresses is sometimes possible (because address A is a justification on one side of the turnstile in one design and address $A^{\perp}$ a justified consequence on the other side in the other design). Once cuts have been done, one development has no addresses available for development, and need to use a pseudo-rule, the "Daimon". It quits the confrontation, the other formula wins and is proved. The positive and the negative designs are then said to converge. Proof is not given by having recourse to exogenous axioms, but by symmetries endogenous to the interactions between one formula and its symmetrical counterpart.

Of course, such a framework is well fitted for language interactions like conversations, maybe better fitted than "dialogic": the structure of the interaction, which needs the contribution of the unproved formulas, is more important than the end of the interaction one formula wins and is proved. In conversations also the structure of the interaction is the main thing, as we do not talk with other people just in order to win a game over our partners. This framework can be adapted to the analysis of a dialog with answers and questions. In this interpretation (Lecomte, Quatrini, Tronçon), the positive rule chooses a question and determines a restricted set of possible answers; the negative rule is used to anticipate all the possible answers that remain offered to the respondent. To develop the proof tree and go up in its development implies to progressively specify the interrogation.

Ludics seems well fitted for analyzing dialogic interactions in conversations. What about speech acts? Since we conceive speech acts as ways of repairing or preventing breakdowns of interactions, we need to know what could be the correspondent of breakdowns of language interaction in ludics.

2 Ludics Breakdowns

Ludics implicitly exclude some breakdowns by its very rules, but it also gives an explicit formal status to other ones. This specificity (in the domain of logic) comes from the idea of confronting the different possible developments of a formula and the developments of its counter-formula. As there are choices to make in the developments (positive steps), only some developments can be related by thread, the thread of a negative Sconse (to propose an empty repertory, forcing the other to abandon by putting the Daimon). Lecomte suggests that we can go a step further. To my thread is associated a promise: if my partner executes my order, then I will myself put the Daimon on that branch of the interaction: the order has been executed, there is no more to demand.

Let us come to the promises. We promise not only because we cannot at the present time make some action, but only in the future, or because we cannot satisfy some conditions at the present time, but only in a longer period (as when we say: I promise to love you). We promise also to counter the possible suspicion of the other that we will not do something or that we do not satisfy some conditions. Promise presupposes that the interaction cannot converge at the present time. The action A is not presently in my ramification. Making $A^{\perp}$ (say, B) would imply another breakdown of the interaction (not Faith, but Bomb). I put myself in a situation that prevents me to play the Bomb. I cannot really block that, but instead I can give to my partner a weapon against the Bomb. Lecomte suggests to use for this purpose the strategy coined by Girard Deterrence. It is the following succession of moves: It starts from a focus on the left of the turnstile the Bomb starts from the right. The following move is negative and offers as a repertory the set of all possible partitions, as in the negative Daimon, but with an exception: the empty set is excluded. As the empty set is the ramification of the Bomb, if the Bomb is played, the Deterrence will conclude to divergence. If on the contrary another accessible ramification is played, the partner is supposed to play the Daimon.

In the promise, I put myself in my partners shoes and offer her in advance the possibility of Deterrence. If Deterrence is possible, if as a consequence Bomb is not played and for this very reason Deterrence is not played, then I will put the Daimon and the interaction will converge on a solution.

Last of all we have to consider excuses. It I have to apologize, I have done something that runs against the desires or the action of another person. I have deprived her from a possible choice. This implies that she was in position of making a choice, that I was in position of offering a choice, and proposed only a repertory that excludes some of her desired actions. I have already put the Sconse in the interaction. The usual answer to the Sconse is the positive Daimon, the other will quit the interaction. But I do not want this disastrous victory. Let us suppose that I could place the other in an inverse situation, the one in which she has the choice, and instead of the negative Sconse, could put the positive Bomb. Then all that I could do would be to converge by putting the negative Daimon. Making excuses is putting the negative Daimon (after having plaid the Sconse), instead of letting the other playing the positive Daimon. I replace my Sconse by a virtual Bomb of her, and as a sign of that I play directly the negative Daimon. If the other accepts my excuses, she can converge with me by playing the positive Daimon, which is the orthogonal of the negative one. The virtual loop will be closed (no real Bomb, in fact), but the positive Daimon of the other people will not have the same bad meaning as at the beginning of the story.

Conclusion

This interpretation of some basic speech acts implies not only that breakdowns are possible in real language interactions as well as in ludics- but also that repairing breakdowns is possible in ludics (or, as we will say, in "para-ludics"). And this implies that we can either replace some past move by another one (as in excuses) or replace in advance a future interactive situation by another one, instead of waiting for development as in proof trees. Logically, this implies a strategy of proof. For speech acts, this implies that one played move compacts several anticipated moves, replacing an anticipated an undesirable sequence by another one, which cannot be justified except by the now counterfactual anticipated sequence. This imply also the possibility to reorganize the alternation between positive and negative moves in ludics, by building blocks that compound several alternate moves (and several ways of alternating moves), as when I imagine myself (as a positive agent) playing at the place of the other as a negative agent, or conversely. Ludics use this compounding device when it aggregates several negative connectors in the same negative move. It could be extended to the analysis of speech acts in a "para-ludic" framework, para-ludic because it deals with repairing or preventing breakdowns in language interactions.

Ludics and Rhetorics

Alain Lecomte[1] and Myriam Quatrini[2]

[1] UMR "Structures Formelles de la Langue", CNRS-Université Paris 8 - Vincennes-Saint-Denis
[2] UMR "Institut de Mathématiques de Luminy", CNRS-Aix-Marseille Université

Abstract. In this paper, we give some illustrations of the expressive power of Ludics with regards to some well known problems often regrouped under the label of Rhetorics. Nevertheless our way of considering Rhetorics encompasses many questions which have been put nowadays in Semantics and Pragmatics.

1 Introduction

Language is mainly interaction. It may be even said, following famous cognitivist S. Pinker [Pinker 07] that *language emerges from human minds interacting with one another*. The main interest of Ludics, for the study of natural language, resides in the possibility it offers for expressing this interaction.

There are many indices of this interaction in language itself and even in syntax, as witnessed by the presence of many small words which have essentially a rhetoric or pragmatic impact, like "*even*", "*nevertheless*" or "*but*". Like the French linguist Oswald Ducrot [Ducrot 1984] said during the eighties, these words cannot be understood simply in truth-conditional terms. "*But*" is not simply "*and*" for instance, and if, for a truth-conditional viewpoint, there is not a big difference between "*few*" and "*a few*", it remains that from a pragmatic side, it is quite obvious that sentences containing these two words cannot be pursued in the same way. If I say *Peter has read few books by Virginia Woolf*, this can be continued by *therefore he does not know her well as a writer*, and if I say *Peter has read a few books by Virginia Woolf*, this can be continued by *therefore he knows her as a writer a little*, and the converse discourse cannot be pronounced. This can be interpreted if we assume that the speaker answers an implicit question which could be : *Does Peter know Virginia Woolf well as a writer?*, and that "*few*" is a negative item, while "*a few*" is a positive one. Vague quantifiers, like "*few*", "*a few*", "*many*", "*a lot*"... denote, as is well known, (approximative) positions on a scale, but specific vague quantifiers have the property of orienting this scale in a direction or in another. "*But*" has a similar property: taking two propositions as inputs, it does not only provide us with a coordination of them, something which can be done by a simple "*and*", it also creates a scale which can be scanned in two opposite directions: one proposition is supposed to be oriented towards one end and the other one towards the other end. This is particularly visible when "*but*" is used to coordinate two quantified expressions, thus requiring they have opposite directions of variation (cf. *he has many relatives but few friends* vs *he has many relatives but many friends*).

Lecomte and S. Trončon (Eds.): PRELUDE 2010, LNAI 6505, pp. 32–57, 2011.

We may also study the phenomenon of presupposition and see about it that it is as if a dialectical structure was at stake. If A says to B : "I don't regret to have been watching the movie", A not only says something about his/her feelings (that s/he does not regret) but also restricts the ways B can react to this assertion because B is supposed to know that A went to see the movie, or if s/he did not know, is required to include in his/her knowledge database that of course A watched the movie. All these facts are well known. They all assume a model of conversation where each speaker takes into account not only the assertions made by the other but also (and perhaps mainly) the *set of expectations* that each speaker has concerning the reactions of the other.

The frame of Ludics, that we will characterize soon, which has been proposed by J. Y. Girard [Girard 01, Girard 03, Girard 06] for foundational purposes with regards to logic, exactly provides a framework where the "actions" of one speaker, seen as "positive", not only depend on the actions of the other of the same polarity, but also on its expected answers, that is "negative" actions.

The body of the paper will present concepts of this framework in a rather intuitive setting, while the Appendix gives more precise definitions. Section 2 presents the ludical notions of *designs* and *normalization*, with their interpretations in proof-theoretic terms and in strategic ones : Ludics formalizes moves in a game as well as in proof search. Section 3 applies these notions to Rhetorics and Eristics. It makes use of an extension of designs (*c-designs*) due to K. Terui [Terui 08] which has two main advantages : it provides a linearized formulation of designs which makes them to look like λ-terms, and it includes *cuts*. Section 4 tries to give an account of *semantics*, as it is generally understood, that is the context-free study of meanings, but instead of staying stuck to a truth-conditional conception of meaning, we envisage it as something built in interactions.

In many respects, our approach has many similarities with a proof-theoretic characterization of meaning, like the one we can find in [Ranta 94]. For this paradigm, meanings are not *truth-conditions*, like in Fregean semantics, but sets of proofs. Applied to Natural Language, and not only to the propositions of Mathematics, we could translate the idea by saying that the meaning of a sentence is the set of facts which prove its validity. For instance, as said by Ranta, the meaning of *Amundsen flight over the North Pole* is provided by... the set of Amundsen's flights over the North Pole! Of course, this set is not made of the "real" facts but of precise indications (localizations) about them, indices which able us to recover the historical facts behind the sentence. But what is missing in the Type-Theoretical approach is the interactive dimension. After all, somebody can object to the truth of the sentence *Amundsen flight over the North Pole*, that is present some other indices which put in doubts the reality of the event denoted by the sentence. This set of indices is itself something we may express in proof-theoretic terms: their proof-theoretic counterpart would be proof attempts to the negation of the sentence, that is something we will call a *counter-proof*. We claim that the meaning of the sentence will reveal during the (generally virtual) *dialogue* between the two attempts, one for proving, the other for disproving, Simplifying things, we say that the meaning of a sentence may be viewed as a set of (potential) justifications.

A's viewpoint is replaced by:

$$\dfrac{\dfrac{0000 \vdash}{\vdash 000}\,(+,0,\{0\})}{< \ > \vdash}\,\mathfrak{C} \qquad \textit{or} \qquad \dfrac{\dfrac{0001 \vdash}{\vdash 000}\,(+,0,1\})}{< \ > \vdash}\,\mathfrak{C}$$

Where $\mathfrak{C}$ is the chronicle $(-,< \ >,\{0\})(+,0,\{\{0\}\})(-,00,\{0\})$.

What we named viewpoints of, respectively S and A, in this example, consists in planning, both for S and for A, what will be their *observable* actions and reactions in a real dialogue. Presupposition assumes they also have *invisible* actions and reactions. In the Appendix, the reader can find the definition of *designs*. Intuitively speaking, designs are sequences of moves, each of which being associated with the application of a rule (positive or negative). Designs are then sequences of steps, some of which are visible and some are not. Computing requires all the "micro"-steps be displayed, even if the observable dialogues don't show them. In this sense, there is a difference between a dialogue (and a strategy in a dialogue) and a "design". In other situations, actions and reactions will stay only *potential* since there is in fact no real dialogue, but only a virtual one. Neverheless, the same situation will occur : designs are foundational objects for dialogues but they are not the dialogues themselves.

2.2 Normalization

Let us examine how the previous *designs*, that we have associated with dialogues, may interact.

1. at the bottom of each design, we have either $\vdash< \ >$ or $< \ >\vdash$, that is twice the same *locus* but with two different polarities. This is exactly similar to a situation where the cut-rule may apply in a sequent calculus. The difference is here that there will be no formulation of *cut* as a rule, but simply we shall consider a cut as a situation where two *loci* meet with two different polarities. In that case, this *cut* may be eliminated according to the standard technique. After the first elimination, what remains is:

$$\dfrac{\dfrac{\dfrac{\dfrac{}{\vdash 0000}\,\dagger \quad \dfrac{}{\vdash 0001}\,\dagger}{000 \vdash}}{\vdash 00} \qquad \vdash 01}{0 \vdash} \qquad\qquad \dfrac{\dfrac{\dfrac{0000 \vdash}{\vdash 000}}{00 \vdash}}{\vdash 0}$$

2. a new cut appears after this first elimination: between $0 \vdash$ and $\vdash 0$. This exactly illustrates the fact that there is a minimal agreement between two speakers: the second agrees to record in his or her own knowledge base the question asked by the first one, what remains is:

$$\dfrac{\dfrac{\dfrac{}{\vdash 0000}\,\dagger \quad \dfrac{}{\vdash 0001}\,\dagger}{000 \vdash}}{\vdash 00} \qquad\qquad \dfrac{\dfrac{0000 \vdash}{\vdash 000}}{00 \vdash}$$

3. again new cuts appear, the first one between $00 \vdash$ and $\vdash 00$, then between $000 \vdash$ and $\vdash 000$, and finally between $0000 \vdash$ and $\vdash 0000$, when this last cut is eliminated, what remains is :

$$\frac{}{\vdash} \dagger$$

something we shall consider a *null* object.

Such a termination will be associated to a *convergence* case (see the Appendix). The two objects which have converged this way when put together are said to be *orthogonal designs*. It is now obvious that our so called third issue does not converge with the presuppositional question made by S since after the first and the second cut-eliminations, there is no cut situation any longer.

2.3 Strategies, Proofs and Designs

One of the main differences with the usual games lies in the fact that there may be no "winner" and no trade-of function in that kind of game : the goal is *not* to win against the other speaker but to reach together a situation in which there is an *agreement* on *expectations*. Such a situation is expressed in terms of convergence. Divergence, on the contrary, may be assigned to *failure*, like *presuppositional failure*.

Let us otherwise notice that, conceived this way, dialectical exchanges seem to occur not by opposing steps to steps, one by one, but by opposing a whole strategy to another one. It is as if each speaker had in his or her own mind, a whole plan, or as if s/he was *projecting* an entire design. This seems to be in agreement with present views in neurosciences, as attested by these words of Jean-Pierre Changeux (himself quoting works by Sperber and Wilson) [Changeux 04, Changeux 09]:

> "Human communication generally takes place in a well defined context of knowledge in which speakers are informing each other [...] Aiming at maximizing the efficience of communication, each speaker tries to recognize and to infer the intention of the one who communicates. In other words, when communication begins, each partner has in his or her own mind the whole possible content of the speech, which constitutes a subset of all his or her knowledge on the world. [...] We may think that each speaker constantly tries to project his or her frame of thought into the mind of his or her co-speaker".

2.4 c-Designs

In place of Girard's designs, we shall mainly use their computational variant due to K. Terui [Terui 08] called *c-designs* (c for "computational"). They extend ordinary designs in many ways. They may contain explicit cuts inside them (contrarily to designs, where cuts are external), and they accept variables as terminative negative subdesigns, thus allowing to represent, in a dialogue, the way in which a speaker gives the flour to another one. Moreover, their linearity and their ability to handle actions not limited to representations in terms of integers make them more legible. Lastly, c-designs allow to define infinite designs by means of their *generators*.

The reader will find a more complete presentation of c-designs in the Appendix. Let us only focalize on their main features. Instead of using the proof-like presentation, the c-designs may be roughly described as generalised λ-terms. In the term calculus which results, we have not a simple, unique application but many ones, in fact as many as there are elements in a signature set $\mathcal{A}$ consisting in a given set of pairs (a, n) where a is a name and n is an arity. The normal terms or cut-free c-designs are still sequences of alternated actions, but :

- positive actions are either constants: $\dagger$ (daimon or abandon) and Ω (divergence or absence of rule), or proper and specific actions (denoted by $\overline{a}$ for a given name a);
- negative actions are either variables $(x, y, z, \ldots)$ or proper negative actions (denoted by $a(x_1, ..., x_n)$).

Then the terms (or c-designs) are defined as follows:

- a negative c-design is either a variable or a sum of negative actions applied with positive c-designs as operands: $a_0(\boldsymbol{x}_{a_0}).P_{a_0} + \cdots + a_k(\boldsymbol{x}_{a_k}).P_{a_k}$;
- a positive c-design is either a constant ($\dagger$ or Ω) or an application which is denoted by $N_0||\overline{a} < N_1, \ldots, N_n >$ where $||$ indicates an interaction (or cut). Such an interaction is an application in the following sense: if N_0 contains a subterm $a(\boldsymbol{x}_a).P_a$ then we have to perform the application $(P_a)N_1 \ldots N_n$. Precisely, in such a case $N_0||\overline{a} < N_1, \ldots, N_n >$ reduces into $P_a[N_1/x_1, \ldots, N_n/x_n]$. Otherwise, if there is no subterm $a(\boldsymbol{x}_a).P_a$ in N_0 (or, equivalentely, if N_0 contains the subterm $a(\boldsymbol{x}_a).\Omega$) the interaction diverges. For instance, the design (i) below is translated into the term (ii):

$$(i)\qquad \dfrac{\dfrac{\dfrac{}{\vdash x.6.1, x.6.2}\dagger \qquad \dfrac{x.6.0.3 \vdash \quad x.6.0.4 \vdash}{\vdash x.6.0}(x.6.0, \{3,4\})}{x.6 \vdash}(x.6, \{\{1,2\},\{0\}\})}{\vdash x}(x, \{6\})$$

$$(ii)\qquad P = x||\overline{\{6\}} < \{1,2\}(x_1, x_2).\dagger + \{0\}(y).y||\overline{\{3,4\}} < \Omega^-, \Omega^- >>$$

where Ω^- is a notation for $\Sigma_{a \in \mathcal{A}} a(\boldsymbol{x}_{\boldsymbol{a}}).\Omega$.

Since in P the first negative term is the variable x, there is no cut. In such a case we use the term "channel" to talk about the variable x and indicates that it is the locus on which an interaction may be plugged in. This design, based on x, consists in a positive action (named $\overline{\{6\}}$) which gives access to two negative c-designs:

- the first one begins by a negative action $\{1,2\}(x_1, x_2)$ which, in principle, binds the variables x_1 and x_2 in any c-design which follows (here $\dagger$, which is a constant).
- the second one performs a negative action $\{0\}(y)$ which binds y inside the positive c-design: $y||\overline{\{3,4\}} < \Omega^-, \Omega^- >$ which follows this negative action. The later positive design reduces to a simple positive action $\overline{\{3,4\}}$ since in fact no (total) negative c-design follows it.

Let us note that c-designs (terms) where $||$ occurs between a mere channel and a subdesign correspond to Girard's designs. Nevertheless, there are cases for which $||$ occurs between two subdesigns : in these cases, c-designs extend Girard's designs, because they now involve *cuts*. $||$ is therefore interpreted as a cut-plugging. For instance, the cut-net (iii) below is translated into the c-design (iv) :

$$(iii)\qquad \dfrac{\dfrac{\dfrac{z.6.1\vdash \quad z.6.2\vdash}{\vdash z.6}}{z\vdash} \qquad\qquad \dfrac{\dfrac{\dfrac{}{\vdash z.6.1, z.6.2}\dagger \quad \dfrac{z.6.0.3\vdash \quad z.6.0.4\vdash}{\vdash z.6.0}}{z.6\vdash}}{\vdash z}}{L \qquad\qquad\qquad\qquad P}$$

$$(iv)\qquad P[L/x] = L||\overline{\{6\}} < N >$$

with:

$$L = \{6\}(z).z||\overline{\{1,2\}} < \Omega^-, \Omega^- >$$
$$N = \{1,2\}(x_1, x_2).\dagger + \{0\}(y).(y||\overline{\{3,4\}} < \Omega^-, \Omega^- >$$

3 Some Applications to the Argumentation Theory

3.1 More on Dialogues

In the previous section, it was seen that by associating a design with a speaker's viewpoint in a dialogue, we may highlight its interactive features. In doing so, we give an account of the interactive content of a discourse : a speech turn is anchored on a locus created by the previous one, due to the other speaker, and creates new *loci* on which the interaction may go on. A dialogue is then just seen as an alternation of speech turns (or utterances for simplicity) on which we may observe:

- the first speech turn : the one which begins the exchange;
- an alternance of speech turns;
- until :
- each intermediate speech turn is anchored on one of the previous speech turns and creates some openings from which the other speaker continue the interaction.

either a turn speech terminates the exchange (because some information has been given, because some agreement has been obtained...)

or the dialogue "fails" on a disagreement, on a misunderstanding feeling ...

EXAMPLE : Let us consider again the small dialogue:

- S: *Were you a smoker ?*
- A: *Yes*
- S: *are you still smoking ?*

Let us take the following notations:

- E , E_1, E_2 and E' respectively denote the utterances *your daughter is dumb because she lost the use of language*, *your daughter is dumb*, *she lost the use of language* and *your daughter lost the use of language because she is dumb*;
- e, e_1, e_2 and e' denote the names of actions which are respectively associated with them.

Let us describe below the c-design $\mathcal{E}$ associated with E:

- $\mathcal{E} = y||\overline{e_1} < \mathcal{N} >$ where:

- y is the channel where an answer may be plugged in;
- $\overline{e_1}$ is the first positive action and corresponds to the claim E_1: *your daughter is dumb*;
- the c-design $\mathcal{N}$ is associated with the justification of E_1 contained in the intervention: *because she lost the use of language*.

- to make $\mathcal{E}$ more explicit, we need to refine the c-design $\mathcal{N}$ which contains as a justification for E_1, the argument E_2. We have : $\mathcal{N} = e_2(x).\mathcal{E}'$. Indeed the negative action $e_2(x)$ gives an account of the position which is ready to support E_2: *she lost the use of language*. Moreover we may forecast[3] that such a support may be the utterance E': *your daughter lost the use of the word because she is dumb*, with which is associated the c-design $\mathcal{E}'$.
- Clearly the c-design $\mathcal{E}'$ is equal to $\mathcal{E}$ except that the positions of e_1 and e_2 are exchanged. Precisely, $\mathcal{E}' = x||\overline{e_2} < \mathcal{N}' >$, where $\mathcal{N}'$ is $e_1(z).\mathcal{E}''$, and $\mathcal{E}'' = \mathcal{E}[z/y]$, and so on ...

Finally the c-design $\mathcal{E}$ associated with the utterance E: *your daughter is dumb because she lost the use of language* is:

$$\mathcal{E}_y = y||\overline{e_1} < e_2(x).x||\overline{e_2} < e_1(z).z||\overline{e_1} \cdots >>$$

The c-design $\mathcal{E}$ is infinite. Nevertheless , we may give a finite presentation of it, by means of a finite generator $\mathcal{G}$ (see in appendix):
$\mathcal{G} = (\{s_1^+, s_2^+\}, \{s_1^-, s_2^-\}, l, s_1^+)$, where the function l is defined as follows:

$$\begin{aligned} l(s_1^+) &= y||\overline{e_1} < s_1^- > \\ l(s_1^-) &= \quad e_2(x).s_2^+ \\ l(s_2^+) &= x||\overline{e_2} < s_2^- > \\ l(s_2^-) &= \quad e_1(y).s_1^+ \end{aligned}$$

[3] And this will be still more convaincing with our semantical formalization: the c-design justifying the utterance e_2 has to belong to the set of c-designs associated with the semantics of e_2. Clearly, the semantics of e_1 and e_2 may share a lot of c-designs, so that the same c-design than the one used to justify e_1 may be reused to justify e_2.

The c-design-like presentation highlights the characteristics of "petitio principii":

- the c-design is infinite and we can say it is a block closed to interaction. Indeed the Adressee can not locate any chronicle: $\overline{e_1} < e_2 \ldots \overline{e_n} < \mathbf{x} > \cdots >$ in order to build some answer as an interacting c-design:

$$e_1 \ldots \overline{e_2} < \ldots e_n \ldots \overline{\mathbf{e_{n+1}}} < \mathcal{M} > >$$

since there is no such variable x.

- the c-design's generator gives an account of the circularity of the argument.

3.2.2 Transfering Premises from a Locus to Another One

A well known work on what is sometimes named *eristic*, according to the ancient Greek word *Eris* meaning "wrangle" or "strife", is the famous *The Art of Always Being Right* written by Schopenhauer. In this book, the German philosopher gives several "stratagems" according to which it is easy to win a controversy against any opponent. For instance, the "fourth stratagem" is the following :

> *Make the opponent to admit the premises of a proposition, in a hidden way during the conversation. Once it is visible that your opponent has conceded all the necessary premises, play the sentence implied by these premises.*

Let us build the c-design associated with a speaker who argues in favour of his or her thesis by refering to premises already accepted by the other speaker.

Let us use the following notations :

- the speaker is referred to by "player" or P
- the addressee by "opponent" or O.
- the utterances corresponding to the premises already accepted by O are denoted A and B.
- we assume that the claim made by P by using the premises already accepted by O, is similar to the following E: *Since A and B (that you accepted) imply C, you will agree that C.*
- we denote by I the utterance *A and B imply C*.
- the names a, b, e, i are respectively associated with the utterances A, B, E and I, and their arities will be made precise below.

We associate with E the following c-design:

$$\mathcal{E} = y || \overline{e_1} < \uparrow (x_a).\mathcal{A}, \uparrow (x_b).\mathcal{B}, \uparrow (x_i).\mathcal{I} >$$

which is built as follows:

- y is the channel where an answer may be plugged in;
- $\overline{e_1}$ is the first positive action : it consists in the claim of the thesis C. This action is ternary since it suggests that the interaction may continue on each element which constitutes the logical argumentation: the two premises and the implication.
- then P is ready to continue the interaction on three channels, represented by three negative actions $\uparrow (x_a), \uparrow (x_b), \uparrow (x_i)$.
- the c-design $\mathcal{I}$ consists in assuming the implication : *A and B imply C*, so it is simply equal to $x_i||\bar{i} < z >$.
- the c-designs $\mathcal{A}$ and $\mathcal{B}$ respectively associated with the support of the premises A and B are supposed to be already built when $\mathcal{E}$ is played, since they have been played previously during the dialogue.

Let us consider the following simplified case: during previous dialogues, the utterances A and B were asserted by P and immediatly accepted by O. We can then represent each of them by some c-design reduced to an elementary positive action, respectively : $x_a||\bar{a} < \Omega^- >$ and $x_b||\bar{b} < \Omega^- >$. We give an account of the transfer of these premises so that they become arguments of the thesis C by the fact that $\mathcal{A}$ and $\mathcal{B}$ are respectively obtained in the following way:

$$\mathcal{A} = \mathcal{F}ax_{x_a}||\bar{a} < \Omega^- > \text{ and } \mathcal{B} = \mathcal{F}ax_{x_b}||\bar{b} < \Omega^- >$$

where $\mathcal{F}ax_y$ is an infinite c-design generated by the following finite[4] generator:

$$(\{s_u\}_{u \in U}, \{s_N\}, l, s_N) \text{ where :}$$

$$l(s_N) = \Sigma_{u \in U} u(\boldsymbol{x}_u).s_u, \quad l(s_u) = y||\bar{u} < s_N, \ldots, s_N > \text{ when } y \notin \boldsymbol{x}_u.$$

Then $\mathcal{F}ax_{y_a}$ is the negative c-design:

$$\Sigma_{u \in U} u(x_1, \ldots, x_n).(y_a||\bar{u} < Fax_{x_1}, \ldots, Fax_{x_n} >)$$

with which the normalisation of a positive c-design $\mathcal{D} = x_a|| \ldots$ gives $\mathcal{D}[y_a/x_a]$.

Finally the c-design $\mathcal{E}$ is as follows:

$$y||\overline{e_1} <\uparrow (x_a).(\mathcal{F}ax_{x_a}||\bar{a} < \Omega^- >), \uparrow (x_b).\mathcal{F}ax_{x_b}||\bar{b} < \Omega^- >, \uparrow (x_i).x_i||\bar{i} < z >>$$

The only possibility for O to successfully continue the dialogue is to use the channel open by the chronicle $\overline{e_1} \uparrow \bar{i} < z >$. If O has nothing to oppose to i, then s/he looses: s/he is obliged to play the daïmon.

Let us underline that such an interaction may be taken into account because we are able to express cuts inside the representations of the speech turns. These internalized cuts allow us to keep the information coming from previous exchanges and to be able to reuse it in the future.

[4] Provided that U (a set of names associated with some utterances) is finite.

4 Ludics and "Semantics"

4.1 On Natural Language Semantics

In the previous sections, we were concerned by *rhetoric*, that is the way in which language is used according to a persuasion goal. Rhetoric is a question of *positions* that speakers occupy in their interlocutory space. In particular, we saw in the previous section how a speaker is bound to make use of the other speaker's expectations. In contrast with rhetoric (which takes place in fact inside what is nowadays called *pragmatics*, that is the study of the *use* of language in context), *semantics* is supposed to deal with the proper content of a sentence (more or less independantly of its context). Traditional Formal Semantics does as if there existed an objectivable sentence meaning which could even be reduced to truth conditions, according to Frege's program. Another viewpoint amounts to consider that :

- meaning is always decided in context, that is, more precisely, in dialectical exchanges,
- meaning is always determined according to reciprocal expectations coming from two speakers (or more) in a dialogue

Of course, when dealing with content, we are obliged to start from some primitive meanings associated with words (*lexical meaning*) and from primitive ways in which those contents may be combined. For instance, when utterring *there is a cat on the mat*, we refer to primitive concepts like those of `cat` and `mat`, and also on relational concepts like `being on`... At a first glance, we may ignore what there is exactly *inside* those concepts! Tarskian semantics would say : `cat` is the concept defined by the function which assigns 1 to every individual x which is a cat, and 0 to the other individuals... We don't think this kind of view bring anything to the comprehension of the semantics of natural language. If we try to reason more in accordance with modern neurocognitive views, we should prefer to say that `cat` is that part of the brain which reacts when the word is heard or when a real cat crosses the road in front of us... But we can also leave the door open to other conceptions : in fact, for us, `cat` will be... a set of designs which defined by the way they interact with other designs at the moment we have an exchange about cats with other people. Of course `cat` as a notion may be more or less deepened in a given situation : it may suffice for us to identify a cat simply by a single feature (its miewing, its whiskers or else...), or we can be in a situation where we expect more, something like a *proof* that there is a true cat! Here the separation theorem is fundamental. It states that designs may be ordered inside the same behaviour. Very long designs may inhabit the behaviour associated with the notion, as well as shorter and more branching ones.

We propose here a conception of *interactive meaning* based on Ludics. Metaphorically, the meaning of a sentence may be viewed by comparison with designs defined by their orthogonal, but this is not only a metaphor, since we are able to give a precise formulation of this idea. The set of designs we associate to a sentence in order to represent its semantics may actually be seen as a set of bearers of potential actions and reactions. It would be possible to give an account of various semantical features and not only of the part which corresponds to logical decomposition. Nevertheless, in this paper, we mainly focus on this part.

Let us take a traditional example of a problem of ambiguity (*scope* ambiguity).

- The following one is still a design that could be convenient :

$$\mathcal{E}_2 = x||\overline{s_1} < \Sigma_{l \in L} q_l(y).(y||\overline{a} < a_1(x_1).(x_1||\overline{\emptyset}), a_2(x_2).(x_2||\overline{g} < \mathcal{N}_1, \mathcal{N}_2) >) >$$

  Here, instead of justifying a_2 by a data, P goes deeper and gives a more detailed justification, for example G: *(the linguist) l spent his childhood in Tunisia and went to a local school*, $\mathcal{N}_1$ and $\mathcal{N}_2$ (not detailed here) are the subdesigns associated with the underlying utterances of G.

Amongst the various designs which may be associated with S (with still the narrow scope for "some") we may observe some of them which share the same successive first actions:

- P asserts s_1. The first positive action is $\overline{s_1}$
- Then s/he is ready to listen objection for every linguist l (the negative actions $q_l(y)$).
- For every linguist l P is able to exhibit some language F, arguing that F is an african language and F speaks l. This is the positive action $\overline{a}$.
- Lastly, if O had still some doubts about one of these two claims (expressed by the negative actions $a_1(x_1)$ and $a_2(x_2)$), P could continue to give justifications, but here, s/he asserts that they are provided by mere datas ($\overline{\emptyset}$).

4.4 Links with the Classical Approach

As a Theory of *Logic*, Ludics allows us to consider also designs as abstractions of proofs, while some sets of designs, provided that they are closed by bi-orthogonality (cf. Appendix), may be seen as formulas (see in the Appendix the notion of *behaviour*). This allows us to view designs as proof attempts. For instance, the design

$$x||\overline{s_1} < \Sigma_{l \in L} q_l(y).y||\overline{a} < \mathcal{A}_1, \mathcal{A}_2 >$$

may be viewed as the following proof attempt in the hypersequentialized polarised linear logic (see the definition of $HSLL$ in Appendix):

$$\dfrac{\mathcal{D}_{l'} \quad \vdots \qquad \dfrac{\dfrac{\mathcal{A}_1}{\vdots}{\downarrow A_1^\perp(F) \vdash} \qquad \dfrac{\mathcal{A}_2}{\vdots}{\downarrow A_2^\perp(l,F) \vdash}}{\vdash \downarrow L(l), \oplus_y(\uparrow A_1(y) \otimes \uparrow A_2(l,y))} \qquad \mathcal{D}_{l''} \quad \vdots}{\dfrac{(\&_x(\uparrow L(x) \multimap \oplus_y(\uparrow A_1(y) \otimes \uparrow A_2(x,y))))^\perp \vdash}{\vdash S}}$$

where S, $L(x)$, $A_1(y)$ and $A_2(x,y)$ are $HSLL$-formulas respectiveley associated with the utterances *Every linguist speaks an african language*, *x is a linguist*, *y is an african language* and *The linguist x speaks y*. And where the designs $\mathcal{D}_{l'}$, $\mathcal{D}_{l''}$, $\mathcal{A}_1$ and $\mathcal{A}_2$ are some proofs attemps of the $HSLL$-sequents[5] $\vdash \downarrow L(l), \oplus_y(\uparrow A_1(y) \otimes \uparrow A_2(l,y))$; $\vdash \downarrow L(l), \oplus_y(\uparrow A_1(y) \otimes \uparrow A_2(l,y))$; $\downarrow A_1^\perp(F) \vdash$ and $\downarrow A_2^\perp(l,F) \vdash$.

[5] Vertical arrows are shift operators - see the Appendix - which make the formulae negative ones, a condition that is necessary if we wish to respect the implicit convention in $HSLL$ according to which formulae are decomposed into maximal blocks of alternate polarities.

REMARKS

1. This interactive meaning analysis enables us to observe that there is no reason to use first order quantifier in the logical decomposition of the meaning of the sentence S. It appears that the convenient logical operations are the additive connectives (extended to a finite set of subformulas indexed by a finite set - here of *linguists* -). Actually, in the foregoing proof attempt, there is no reason that the subdesigns $\mathcal{D}_l$, $\mathcal{D}_{l'}$, $\mathcal{D}_{l''}$ would be based on similar attempts to find a proof. That would be a necessity if we used the universal quantifier, as shown in [Fleury-Quatrini 04].

2. For the sake of simplicity, we have not dealt with the part *l is a linguist*. Then in the foregoing proof scheme the formula $\uparrow L(l)$ is simply weakened. It is of course possible to refine the analysis by taking into account possible exchanges around the fact that such or such l is indeed a linguist. And thus to recover more detailed proof schemes.

Finally, we retrieve as a particular case of interactive meaning, the usual notion of "logical form" and among its consequences, the treatment of quantifier scope. We may of course also retrieve the notion of truth value. Actually, provided that it contains a "winning" design[6] a behaviour may be seen as a truth formula.

We may then consider, inside the set of designs S^*, a subset which is a behaviour which corresponds to the $HSLL$ formula (still denoted by S) :

$$S = (\&_x \uparrow L(x) \multimap (\oplus_y \uparrow A(y) \otimes \uparrow P(x,y)))) \oplus \oplus_y \uparrow A(y) \otimes (\uparrow L(x) \multimap \&_x \uparrow P(x,y))).$$

This behaviour is itself the combination of more elementary ones by operations it is possible to define on behaviours which exactly reflect the usual connectives of linear logic!

Nevertheless, we insist upon the fact that logical meaning so characterized is *only* a subpart of the sentence semantics. We may also deal with pragmatical aspects of semantics in a way we outlined in section 3.

5 Conclusion

Intuitively speaking, Ludics allows us to develop a new viewpoint, according to which very various objects may be assigned to sentence (and word?) meaning. These objects are technically characterized as *sets of designs*, among them some *stable* sets (with regards to bi-orthogonality) give so-called *b*ehaviours. We don't need to know what is the *essence* of meaning because those objects are defined by their reactions with regards to other ones they are interacting with. As pointed out by C. Faggian [Faggian 2006], only the properties of these objects that can be tested by means of interaction with objects of the same kind can be observed, they are the *observables*.

Moreover, we may make the depth of the characterization vary, according to a *separation* theorem.

[6] Roughly speaking a winning design is a design which does not use any daimon. It is, for example, the case of our design $\mathcal{E}_0$.

Let us finally emphasize the following points:

- In Ludics, neither formulae nor the sets of designs (behaviours) correspondng to them are the primitive objects. The meaning of a sentence is thus assigned an object which is not *a priori* closed : the meaning of a sentence may be *more and more refined*. In particular, the order between designs given by the separation theorem enables one to explore more and more precisely the argumentative potential of the sentence.
- Ludics is built with an explicit attention given to the "logical frontier" (what falls inside logic *versus* what falls not). Logical concepts like formulae, proofs and connectives are defined in a world which is larger than the strict logical word (let us remember that we have *paralogisms* like the *daïmon*, and counter-proofs in that world!). This feature may be used to formalize aspects of meaning which don't directly deal with Logic, like it is the case in pragmatics, dialectic and rhetoric, as seen in the first part of this paper.
- Ludics also refers to the possibility of playing on various interpretations of logical concepts : *localized* vs *delocalized* (or *spiritualist* in Girard's sense), where we see an implementation of the well known distinction between *tokens* and *types* (see [Strawson 50]). The same sentence can be viewed (inside the same framework) as a token - when seen as an utterance made in a given context - as well as a type - when seen as delocalized and understood independently of any context. Normalization with $\mathcal{F}ax$ makes the communication possible between the two.
- Ludics also enables us to deal with *dynamics* for free, like we saw it in the first part of this paper. This feature is particularly highlighted in the use of c-designs, which include *cuts* and therefore a procedure of normalization which is similar to β-reduction.

References

[Andréoli 92] Andréoli, J.-M.: Logic Programming with Focusing Proofs in Linear Logic. Logic Programming with Focusing Proofs in Linear Logic 2(3), 297–347 (1992)

[Aristotle] Aristote Les Réfutations Sophistiques, trad. Jules Tricot, ed. Vrin, Paris (1995)

[Changeux 04] Changeux, J.-P.: L'homme de vérité. In: Jacob, O. (ed.), Paris (2004)

[Changeux 09] Changeux, J.-P.: Du vrai, du beau, du bien. In: Jacob, O. (ed.), Paris (2009)

[Curien 2004] Curien, P.-L.: Introduction to linear logic and ludics, part I and II, to appear, downloadable, from `http://www.pps.jussieu.fr/ curien/LL-ludintroI.pdf`

[Ducrot 1984] Ducrot, O.: Le dire et le dit, Editions de Minuit, Paris (1984)

[Faggian 2006] Faggian, C.: Ludics and interactive observability: the geometry of tests. Theoretical Computer Science 350(2), 213–233 (2006)

[Fleury-Quatrini 04] Fleury, M.-R., Quatrini, M.: First order in Ludics. Mathematical Structures in Computer Science 14(2), 189–213

[Girard 99] Girard, J.-Y.: On the Meaning of Logical Rules-I. In: Berger, U., Schwichtenberg, H. (eds.) Computational Logic, Springer, Heidelberg (1999)

[Girard 01] Girard, J.-Y.: Locus Solum. Mathematical Structures in Computer Science 11, 301–506 (2001)

[Girard 03] Girard, J.-Y.: From Foundations to Ludics. Bulletin of Symbolic Logic 09, 131–168 (2003)

[Girard 06] Girard, J.-Y.: Le Point Aveugle, vol. I, II. Hermann, Paris (2006)

[Hamblin 70] Hamblin, C.-L.: Fallacies. Vale Press, Newport News, republished in (2004)

[Hintikka-Kulas 83] Hintikka, J., Kulas, J.: The Game of Language: Studies in Game Theoretical Semantics and its Applications. D. Reidel, Dordrecht (1983)

[Hintikka-Sandu 97] Hintikka, J., Sandu, G.: Game Theoretical Semantics. In: Van Benthem, J., ter Meulen, A. (eds.) Handbook of Logic and Language, ch. 6. Elsevier, Amsterdam (1997)

[Martin-Löf 84] Martin-Löf, P.: Intuitionistic Type Theory. Bibliopolis, Naples (1984)

[Pietarinen 07] Pietarinen, A.-V.: Game Theory and Linguistic Meaning. Elsevier, Amsterdam (2007)

[Pinker 07] Pinker, S.: The Stuff of Thought, Language as a Window into the Human Nature. Penguin Books (2007)

[Ranta 94] Ranta, A.: Type-Theoretical Grammar. Oxford University Press, Oxford (1994)

[Schopenhauer] Schopenhauer, A.: The Art of Always Being Right. Gibson Square Books Limited (2004)

[Strawson 50] Strawson, P.F.: On Referring. Mind 59 (1959)

[Sundholm 86] Sundholm, G.: Proof Theory and Meaning. In: Gabbay, D., Guenthner, F. (eds.) Handbook of Philosophical Logic, vol. III, pp. 471–506. D. Reidel, Dordrecht (1986)

[Terui 08] Terui, K.: Computational Ludics, to appear in Theoretical Computer Science, J.-Y.Girards Festschrift special issue (2008)

[Tronçon 06] Tronçon, S.: Dynamique des démonstrations et théorie de l'interaction, PhD thesis, Université d'Aix-Marseille (2006)

[Wittgenstein 53] Wittgenstein, L.: Philosophische Untersuchungen. Blackwell, Malden (1953)

Appendix A: A Very Short Presentation of Ludics

Ludics in a Nutshell

Ludics is a recent theory of Logic introduced by J.-Y. Girard in [Girard 01]. Here we don't give the entire definitions of the core concepts of Ludics but we just give an account of the objects we use in this paper.

Designs

At a first glance, *designs* look like *proofs*. In fact, they come from a deep study of proofs and their interaction. It was discovered already during the nineties that proofs of Linear Logic could be decomposed into blocks of opposite polarities (positive like for $\otimes$ and $\oplus$ steps, negative like for $\&$ and $\wp$ steps). That opened the field to a polarized and focalized logic. In such a frame, blocks of a given polarity are reduced to only one step corresponds to the positive one. In the later case, the process fails (or *diverges*). The process may not terminate since designs are not necessarily finite objects.

When the normalization between two designs $\mathcal{D}$ and $\mathcal{E}$ (respectively based on $\vdash \xi$ and $\xi \vdash$) succeeds, the designs are said to be *orthogonal*, and we note: $\mathcal{D} \perp \mathcal{E}$. In this case, normalization ends up on the particular design :

$$\frac{}{\vdash}\dagger$$

Let $\mathcal{D}$ be a design, $\mathcal{D}^{\perp}$ denotes the set of all its orthogonal designs. It is then possible to compare two designs according to their counter-designs. Moreover the separation theorem [Girard 01] ensures that a design is exactly defined by its orthogonal: if $\mathcal{D}^{\perp} = \mathcal{E}^{\perp}$ then $\mathcal{D} = \mathcal{E}$.

Behaviours
One of the main virtues of this "deconstruction" is to help us rebuilding Logic.

- Formulae are now some sets of designs. They are exactly those which are closed (or stable) by interaction, that is those which are equal to their *bi-orthogonal*. Technically, they are called *behaviours*.
- The usual connectives of Linear Logic are then recoverable, with the very nice property of *internal completeness*. That is : the bi-closure is useless for all linear connectives. For instance, every design in a behaviour $\mathbf{C} \oplus \mathbf{D}$ may be obtained by taking either a design in $\mathbf{C}$ or a design in $\mathbf{D}$.
- Finally, *proofs* will be now designs satisfiying some properties, in particular that of not using the daïmon rule.

The c-Designs

In [Terui 08] K. Terui proposes an alternative formulation of Ludics which is motivated by stakes of "developing a monistic, logical and interactive theory for computability and complexity". In order to follow such a program, K. Terui modifies and extends the formalism for Ludics.

We focus here on the notions of c-designs and generators that we use in our text and we propose a very simplified presentation of them.

c-Designs
Among the new features of the c-designs compared to the original ones of Girard let us underline the following:

- Instead of objects with absolute address, the c-designs may be described using a term calculus approach. The absolute addresses are replaced by variable binding.
- The c-designs extend ordinary designs in that they contain explicit interactions.

We then focus on some technical modification into the designs building. The c-designs still contain sequences of alternated actions, but we may at first observe that we have a new notion of action. The c-designs are defined according to a signature set $\mathcal{A}$: a set of couples (a, n) where a is a name and n is its arity. And the positive actions are either constants: $\dagger$ (daïmon or abandon) and Ω (divergence or absence of positive rule), or proper and specific actions (denoted by $\overline{a}$ for a given name a) while the negative actions are either variables (x, y, z,...) or proper negative actions (denoted by $a(x_1, ..., x_n)$). Secondly, designs contain also cuts which enables to consider applications in a term calculus approach. Let us underline that in such a term calculus , we do not have a unique application but as many applications as elements in a signature set $\mathcal{A}$. Then the terms or c-designs are co-inductivey defined:

- The positive c-designs are: $P = \quad \Omega \quad | \quad \dagger \quad | \quad N_0||\overline{a} < N_1, \ldots, N_n >$
- The negative c-designs are: $N = \quad x \quad | \quad \Sigma_{a\in\mathcal{A}} a(\boldsymbol{x}).P_a$

The positive designs really containing a cut are designs $N_0||\overline{a} < N_1, \ldots, N_n >$ when N_0 is not a variable. In such a case the cut may be seen as an application in the following sense: if N_0 contains a subterm $a(\boldsymbol{x}_a).P_a$ then we have to perform the application $(P_a)N_1 \ldots N_n$. Precisely, in such a case $N_0||\overline{a} < N_1, \ldots, N_n >$ reduces into $P_a[N_1/x_1, \ldots, N_n/x_n]$. Otherwise, if there is no subterm $a(\boldsymbol{x}_a).P_a$ in N_0 (or, equivalentely, if N_0 contains the subterm $a(\boldsymbol{x}_a).\Omega$) the interaction diverges.

When the negative subdesigns are all variables the c-design is said to be a cut-free design.

Let give as an instance of c-design the one corresponding to the $\mathcal{F}ax$. It is a negative c-design recursively defined as follows:

$$Fax_y = \Sigma_{a\in\mathcal{A}} a(x_1, \ldots, x_n).(y||\overline{a} < Fax_{x_1}, \ldots, Fax_{x_n} >)$$

Generators

K. Terui introduces in [Terui 08] design generators that provide a means to finitely describe infinite designs.

A **generator** is a triple (S^+, S^-, l) where S^+ and S^- are disjoint sets of states and l is a function defined on $S = S^+ \cup S^-$ satisfying the following conditions:

- For $s^+ \in S^+$, $l(s^+)$ is either Ω, $\dagger$ or an expression of the form $s_0^-||\overline{a} < s_1^-, \ldots, s_n^- >$ such that the s_i^-'s belong to S^-.
- For $s^- \in S^-$, $l(s^-)$ is either a variable x, or an expression on the form $\Sigma_{a\in\mathcal{A}} a(\boldsymbol{x}).s_a^+$ such that the s_a^+'s belong to S^+.

A **pointed generator** is a quadruple (S^+, S^-, l, s_I) where (S^+, S^-, l) is a generator and $s_I \in S$.

We say that (S^+, S^-, l, s_I) generates a c-design called $design(S^+, S^-, l, s_I)$.

A c-design D is **finitely generated** if it is generated by a pointed generator which has finitely many states, and whenever $l(s^-) = \Sigma_{a\in\mathcal{A}} a(\boldsymbol{x}).s_a$, all but finitely many s_a have the label Ω.

Examples:

- the pointed generator $(\{s_\dagger\}, \{s\}, l, s_\dagger)$, with: $l(s_\dagger) = \dagger, \quad l(s) = \Sigma_{a\in\mathcal{A}}.s_\dagger$ generates the negative daïmon: $\Sigma_{a\in\mathcal{A}}.\dagger$.
- the pointed generator$(\{s_a\}_{a\in\mathcal{A}}, \{s_N\}, l, s_N)$ with:

$$l(s_N) = \Sigma_{a\in\mathcal{A}}(\boldsymbol{x}_a).s_a \text{ and } l(s_a) = y||\overline{a} < s_N, \ldots, s_N > \text{ if } y \notin \boldsymbol{x}_a$$

generates the $\mathcal{F}ax$.

Remark: Provided that $\mathcal{A}$ is finite Dai^- and $\mathcal{F}ax$ are finitely generated.

Appendix B: An Hypersequentialized Linear Calculus

We give here a short presentation of a hypersequentialized version of linear calculus, which enables one to manipule the designs as (para)proofs of a logical calculus.

Formulae and Sequents

By means of polarity, we may simplify the calculus by keeping *only positive formulae*. Of course, there are still negative formulae... but they are simply put on the left-hand side after they have been changed into their negation. Moreover, in order to make para-proofs to look like sequences of alternate steps (like it is the case in ordinary games), we will make blocks of positive and of negative formulae in such a way that each one is introduced in only one step, thus necessarily using *synthetic connectives*. Such connectives are still denoted $\oplus$ and $\otimes$ but are of various arities. We will distinguish the case where both $\oplus$ and $\otimes$ are of arity 1 and denote it $\downarrow$.

- The only linear formulae which are considered in such a sequent calculus are built from the set P of linear constants and propositionnal variables according to the following schema :

$$F = P|(F^{\perp} \otimes \cdots \otimes F^{\perp}) \oplus \cdots \oplus (F^{\perp} \otimes \cdots \otimes F^{\perp})| \downarrow F^{\perp}$$

- The sequents are **denoted** $\Gamma \vdash \Delta$ where Δ is a multiset of formulae and Γ contains at most a formula.

Rules

- There are some axioms (logical and non logical axioms):

$$\overline{P \vdash P} \qquad \overline{\vdash 1} \qquad \overline{\vdash \downarrow T, \Delta} \qquad \overline{\vdash \Delta}^{\dagger}$$

  where P is a propositionnal variable ; 1 and T are the usual linear constants (respectively positive and negative).

- The "logical" rules are the following ones :

  Negative rule

$$\frac{\vdash A_{11}, \ldots, A_{1n_1}, \Gamma \quad \ldots \quad \vdash A_{p1}, \ldots, A_{pn_p}, \Gamma}{(A_{11}^{\perp} \otimes \cdots \otimes A_{1n_1}^{\perp}) \oplus \cdots \oplus (A_{p1}^{\perp} \otimes \cdots \otimes A_{pn_p}^{\perp}) \vdash \Gamma}$$

  Positive rule

$$\frac{A_{i1} \vdash \Gamma_1 \quad \ldots A_{in_i} \vdash \Gamma_p}{\vdash (A_{11}^{\perp} \otimes \cdots \otimes A_{1n_1}^{\perp}) \oplus \cdots \oplus (A_{p1}^{\perp} \otimes \cdots \otimes A_{pn_p}^{\perp}), \Gamma}$$

  where $\cup \Gamma_k \subset \Gamma$ and for $k, l \in \{1, \ldots p\}$ the $\Gamma_k \cap \Gamma_l = \emptyset$.

Remarks on Shifts

Using the shift is a way to break a block of a given polarity. Separate steps may be enforced by using the *shift* operators $\downarrow$ and $\uparrow$ which change the negative (resp. positive) polarity into the positive (resp. negative) one. The rules introducing such shifted formulae are particular cases of the positive and the negative one:

$$\frac{A^{\perp} \vdash \Gamma}{\vdash \downarrow A, \Gamma}[+] \qquad\qquad \frac{\vdash A^{\perp}, \Gamma}{\downarrow A \vdash \Gamma}[-]$$

where A is a negative formula.

Example. In a block like $A \otimes B \otimes C$ in principle, A, B and C are negative, but if we don't want to deal with A, B, C simultaneously, we may change the polarity of $B \otimes C$ (which is positive) and make it negative by means of $\uparrow$. We write then $A \otimes \uparrow (B \otimes C)$.

Compare the two following partial proofs, where (1) does not use any shifts and (2) uses one :

instead of (1): $\dfrac{A^{\perp} \vdash \quad B^{\perp} \vdash \quad C^{\perp} \vdash}{\vdash A \otimes B \otimes C}$ we get (2) : $\dfrac{A^{\perp} \vdash \quad \dfrac{\dfrac{B^{\perp} \vdash \quad C^{\perp} \vdash}{\vdash B \otimes C}}{\downarrow (B \otimes C)^{\perp} \vdash}}{\vdash A \otimes \uparrow (B \otimes C)}$

2 Web Sites: Which Situation ?

2.1 What Is Behind the Screen ?

The Web usage evolved since the first static web pages developed by Tim Berners-Lee for the CERN around 1990, that initiated the World Wide Web as we know it now. The http protocol created for that purpose has not really been modified: a request is sent to a server containing a page to be uploaded, the server sends back the html page in case it exists. Behind the screen, the situation is more complex. It involves now the following procedures: dynamic generation of pages, code distribution between client and server, multiple requests between client and server to complete the download, use of the hierarchical structure of a page. A *page*, as it is viewed by a user, contains *static data* and *hyperlinks*, shortly called links later, organized hierarchically in *frames*. The html protocol is responsible for specifying how data should be displayed by the browser and how requests may be built from available hyperlinks. Data may be of any kind of media: texts, images, music, videos,[1] and so on. Links are nothing else but (authorized) server requests. They precise the address to be called and values given to parameters if needed. A request is built from this URL, the operation to be used, e.g. GET or POST, and a body containing data (date, character encoding, and other pieces of information). Parameter values of the request may be partially or totally hidden to the user when viewing the request. With a GET form, parameters are sent as part of the URL as in `http://www.optima.fr/proust/type_list.php?type=notebook` where `notebook` is the value of parameter `type`. With a POST request, parameter values are included in the body of the request, hence are no more immediately visible to the user. In the following, we do not care about these technical differences.

Even if static pages still exist, a large majority of pages are now dynamically generated either by the server or the client. Web sites are now programmed in html-embedded languages, e.g. *php*, *asp*, *jsp*, or in general-purposes language, e.g. *Perl*, *Java*, or any other standard programming languages where data are transmitted and received by means of the CGI protocol (Common Gateway Interface). Servlets, applets, or other technologies, as well as dedicated libraries or environments (.NET, J2EE, ...), help the programmer in developping her web sites. Note that some programming languages are interpreters and almost all web sites are conceived without any possible analysis of *html* and/or *xhtml* data that may be generated. Code able to answer requests may be downloaded in an initial request from the server to the client browser, just to speed up the computation. This is the case, for example, with *JavaScript*. The dynamic generation consists in constructing the data to be included in the page together with links the server can answer to (as we hope if the page is coherent with respect to what the server is able to do). It may involve from the server requests to other servers.

Generated pages are hierarchically structured into *frames*. The reason is twofold. First, structuring the page eases its development by increasing its modularity and reusability. Second, transferring data from the server to the client may become slow as data volume increases (e.g. with video), even though requests from the user concern only small changes in the initial page. For example, Ajax mechanism uses frame

[1] Even though sounds and videos seem to be dynamic for the user: we abstract from the computation their listening or viewing requires.

replacement instead of whole page download and relies on a tree specification of the page. It supposes the possibility to index each frame as a path in the tree.

Without any significant changes since the beginning of the World Wide Web, the *http* protocol that supports the infrastructure of web interaction is *stateless*, i.e. exchanges between server and client are not correlated. However web is typically used as a sequence of interactions between a user and a server: the user sends a request, the server computes a response following the *html* protocol, that may include links, the user clicks on one of these links sending a new request, and the process goes on until the user has her answer, grows tired of browsing or there is no more available links. The server being called by an indefinite number of users has to add a piece of information in order to distinguish this dialogue from others pending interactions. This piece of information allows the server to recover the context and the history (maybe stored in databases on the server side). Such a piece of information is standardly written as a session identifier systematically put in the request, or as a cookie stored on the client side if information should be kept through different sessions of the same user. Remark that situation is worse in most programming implementations: for efficiency purposes, the server launches a new thread (with an a priori fresh environment) each time it receives a request, the thread dies whenever the answer is sent back, whence throwing away the state of the environment. On the opposite, the execution of standard programs, say your favorite software, behaves differently: a thread is also started with a fresh environment and do not quit before the end of the computation (that may include reads and writes). What is at stake there ? 'Addresses' of computation. In fact, a run of a software goes through various stages characterized by an 'address' given as the memory address of the next instruction to be executed in the program and the content of the heap (not to speak of the memory itself). Such addresses are still available during reads and writes but not as soon as the thread dies. Cookies and session identifiers are nothing else but a way to restore an environment and the current address of execution in that environment. In the following, we shall see that address usage is a key ingredient of Ludics. Concretely, the programming language we present in Section 4 keeps closures and objects created during interactions in the current environment, the addresses of which are part of URL calls. State restoration is a particularly difficult task when user 'plays' with the history of her navigation. In fact, from the beginning of web site creation, back, reload, refresh and cloning operations are available in menus of browsers or as specific links present in pages. Cloning allows to duplicate the current page (in a new window or tab) possibly without any request to the server. This may confuse the server as the cookie and the session identifier may come with the same values from two different windows, although the session should have been split as soon as cloning occurs. The reload operation sends the last executed request to the server and displays its result, contrarily to a refresh operation that only prints again what was already printed without any further request (except explicitly mentioned). The back operation sends the penultimate request. Finally, remark care has to be taken in case of reservation, payment, ... as repetition of requests may induce multiple results. However, this is a tedious task for the programmer as recovering of environments may be done as many times as desired: a resource-conscious model may be more accurate. All these operations are problematic as their semantics is not fully part of the current models of the web languages.

2.2 Drawbacks of Current Programming Environments

Three other peculiarities have to be mentioned: asynchronicity, run-time error management and correctness of dynamic generation of an html result. Contrarily to standard programming, where an expression, e.g. a function call, is evaluated as soon as its value is requested, a user request may be indefinitely delayed because of network overload, data losses or other problems. Such asynchronicity is not fairly taken into account and the only method in use concerns a time threshold 'wired' at the operational level that raises some sort of exception. Managing run-time errors poses two different kinds of problems. Remember that *types* finitely abstract data structures and functions. In quite all programming languages, a *type checker* controls statically if possible the validity of function calls. This is particularly true with fully-typed functional languages, hence free from execution errors. In the case of web, data are transferred as character strings without any distinction even between basic types. Typing is then more difficult to test and should be done largely by means of a dynamic checker. The second kind of problem is more relevant to web services than web sites. It concerns the following of the interaction: the result of a request is another set of possible requests, and so on. Checking types is no more reduced to the first interaction but to the full dialogue until one of the two client or server stops the exchange. This means that types associated to web links should be tree-like structures of actions instead of just one operation resulting generally in a basic data type.[2] This is particularly true with web services. When dealing with web services, server and client are programs and it is essential that interaction goes on without errors. In the community of web services, a *contract* is a service description document that characterizes an expected interaction. Server and client must adhere collectively to a shared communication agreement. There does not exist currently a general standard language for that purpose, even if a well-defined type language may really be the good solution. Finally, the last particularity of web concerns the correctness of an html page dynamically generated. html documents are part of so-called semi-structured data: their content may vary accordingly to schemas. Schemas are specifications that include optional elements, repetitions, and so on. Because of that, their typing is intrinsically more difficult than with usual data structures. If this was not enough, most programming languages used in web design, in order to favor modularity, accept that functions or procedures generate strings, untypable as html data, hoping that the final result sent to the user conforms to some schema. As this is clearly outside the scope of this paper, we do not develop this point. Although this is taken into account in the concrete language FICX presented later.

Let us summarize in other words characteristics of web sites sketched above. A client interacts with a server by requesting maybe part of a tree of html data. To prevent the user from trying blindly allowed continuations, the server presents them before via the browser to the user as a set of available links. Data surrounding these links give "explanations" on their usage, or just final information ending the dialogue between the server and the client. Each request is a call with arguments sent with the request or retrieved by subsequent server calls. Particular operations are available to travel through the history of the interaction. However these operations are frequently sources of problems as they

[2] Obviously, a result may be a function in functional languages, or pointers to functions in imperative languages.

may induce cycle behaviours. Finally, types to be used have to include an abstraction of the full process. Taking care of the necessity of a full type checking leads to consider a functional language as the underlying programming paradigm. The fact that web interaction behaves as an alternance of question and answer motivates the use of Ludics as a type language.

2.3 Concepts in Use in Related Works Coming from the Functional Paradigm

If we abstract from technologies put forward, a few concepts emerge in the functional programming community to cope with web particularities. *Continuations* and *monads* are the most important ones. Queinnec [8, 9] was the first that motivated the continuation principle as an essential feature able to give an operational semantics to web traits such as back and reload operations. Developing a CD-ROM to be used by students, Queinnec noticed that "the use of a browser to compute over the web naturally requires the language, in which computations are expressed, to support continuations and concurrency" [8]. Continuations are used in the following way: whenever a page is *shown* to the user via a browser (call to a `show` function, Fig. 1 on the left), a continuation is built (call to `register-continuation`), the name of which is passed into the content of the web page as part of URLs to be requested. The continuation is a function containing the code to be executed if the user asked for this link. A request to the server, e.g. `http://my.scheme/resume?contine=k87&exp=4`, includes the continuation name `k87` together with parameter values if required (`exp=4` in the example). It is the server responsability to retrieve the correct function to be called (call to `get-registered-continuation` in function `server`, Fig. 1 on the right).

Thiemann [11, 12] advocated for the use of monads to design a CGI mechanism. A CGI program is dedicated to parse requests and send html pages back to the client. Furthermore, each CGI call induces a new thread of computation that ends when answer is sent. By the way, there is no implicit notion of *session* inside CGI, where a session is the sequence of interactions between a specific user and a specific site. As already mentioned, a standard way consists in logging previous data and operations and use session identifiers inserted in the URL or cookies in the body of a request to recover a previous state. Monads [13] may serve that purpose. They are intended to represent computations, i.e. procedures possibly with data. Thiemann encapsulates in a monad the current state of the environment together with the function associated to the request. He uses it also to give a better modularity and control on the construction of html pages. The implementation proposed by Thiemann imposes the state to be regenerated (by

```
(define (show page)
  (call/cc (lambda (resume)
    (let (
        (url (register-continuation! resume))
        (connection (current-connection))
        )
      (display (page url) connection)
      (close-output-port connection)
      (suicide)
) ) )       )
```

```
(define (server request)
  (let* (
      (c (get-param request "continue"))
      (k (get-registered-continuation c))
      )
    (if k
      (k request)
      ...)
) )
```

Fig. 1. Example of continuations use in web programming [8]

reexecuting scripts) on each request, but this may be omitted if a closure of the memory is done instead of logging only data and operations.

In his seminal paper dedicated to *Arrows* [7] in the context of functional programming, Hughes presented an original way of computing CGI programs for Web pages. The *Arrow* concept generalizes the monadic one in such a way that a program may be viewed as a plug of functions taking one or several inputs and outputs. Such functions are lifted to arrows objects and combined by means of specific operators. The following code examplifies these aspects:

```
arr (λz → "What is your question?") >>> ask
   >>> (arr id &&& ask)
   >>> arr(λ(q,a) → "The answer to \"" . q . "\" is " . a)
```

The arrow function `arr` transforms a function into a computation, i.e. an object to be combined with other arrows. The operation >>> composes two arrow computations (hence as if a function name followed an argument) while **&** builds a pair from the results of two arrows. `ask` sends its argument (written before `ask` in the code) to a client while the answer is the result of the call. The previous code sends the sentence `"What is your question?"` to the user whose answer is supposed to be a question sent back as is (second call to `ask`). If we suppose the question from the user is "`How old are you ?`" and the user's response to this question is "`40`", these two elements are passed as the two arguments of the last unnamed function that sends back to the client the sentence "`The answer to "How old are you ?" is 40`". The shift from monads to arrows allows to better limit the elements of the environment that are necessary for resuming a computation after an `ask` operation. Moreover, even if a monad is generic and simple to use, the overall structure of a program using monads remains a directed acyclic graph of computations. Arrows usage frees the programmer from this constraint and allows for a more efficient program execution.

Using monads or arrows eases the programmer task by providing the necessary tools for high-level and modular code development. However these models do not integrate a full and coherent semantics for back and reload operations.

Monads or arrows as well as continuations give concrete ways to implement web sites inside a functional language. It ensures that a strict typing of web programs is possible, even if interaction between a user and a server has to be dynamically type checked. Moreover, a verification may be done between hyperlinks provided in a page and continuations, monad or arrow evaluations available in the server side: at least, the server should answer to the links it offers ! However these models do not fully integrate the dialogue itself as a fundamental process. Using dialogue as the core mechanism, its extension to web services should be made easier. The language we propose in Section 4 has as peculiarities:

- a functional language as a basis programming paradigm to ensure type checking at a low level,
- a reaction model, that may be implemented by means of continuations, monads or arrows, to take into account asynchronicity,
- a fully-typed programming language able to check sessions and not local interactions,
- a modular system close to the frame model in use by web designers.

3 Web and Ludics

This section is devoted to an interpretation of web dialogue in terms of Ludics. We first present briefly Ludics and two of its principal ingredients, *designs* and *behaviours*. We recall the definition of a *cut* between two designs. We show then in which extent designs may model user and server processes whereas web interaction is nothing else but a cut reduction. In particular, we recover concepts mentioned in the previous section and specifically the importance of the notion of address. Later on, we use Terui's works [10] to reformulate Ludics in such a way that it corresponds closely to the type system involved in our programming language (excepted modular aspects and other technicalities).

3.1 Interaction in Ludics

J.-Y. Girard defines *Ludics* [5, 6] as a pre-logical framework upon which (linear) logic is (re)built. Linear Logic (LL,[4]), also defined by J.-Y. Girard, is deduced from a decomposition of classical implication: $A \rightarrow B = !A \multimap B$ meaning that a resource A has to be reusable (so the ! modality) for being an argument to a function using each resource linearly (so the change of implication from classical $\rightarrow$ to linear $\multimap$), the function producing a resource B. This change of viewpoint allows for a nice semantics for logical proofs and new ways of presenting proofs. By the way, classical 'and' and 'or' connectives are each decomposed into multiplicative ($\otimes$ 'and' and $\parr$ 'or') and additive ($\&$ 'and' and $\oplus$ 'or') versions together with specific *exponential* modalities to treat reusability. Notice that it gives also a profound insight on duality by extending intuitionnistic logic. *Ludics* is an attempt to reconstruct Linear Logic (but without its exponential part) from the viewpoint of duality: (i) the meaning of a formula is a set of structures called *designs*, (ii) a notion of duality between designs induces duality between formulas, and (iii) the space of formulas is fully describable in terms of multiplicative and additive connectives over formulas. The key ingredient to specify what kind of structure should be given to designs comes from Andreoli's works on focalization w.r.t. logical programming [1]: a proof can be organized in such a way that decomposition of clustered deterministic (say negative) and non-deterministic (say positive) connectives in a formula alternate. Furthermore, although every negative formulas[3] may be concurrently and immediately decomposed, one positive formula may be chosen when a positive step occurs. Such a positive formula is called a *focus*. Following focalization, we can consider that sequents has at most one negative formula. A decomposition step, i.e. a bottom-up application of a rule in a sequent calculus, consists in

- either choosing a positive formula to be decomposed, to give raise to a set of (negative) subformulas, hence a set of sequents (one for each negative formula),
- or decomposing the negative formula to give raise to a set of sets of (positive) subformulas, hence a set of sequents (one for each set of positive subformulas).

Designs may be viewed as abstracting concrete and focalized proofs, and taking into account infinity and failures. With that in mind, one throws away (at that moment)

[3] A negative formula has a negative connective as the main one.

the notion of formula to only keep the one of **locus**, or *address*: a subaddress specifies a subformula when an address denotes a formula. The subaddress of an address is nothing else but the address completed by some index, e.g. an integer, called a *bias*.

Definition 1. *An* **action** *abstracts from a decomposition step (part of for the treatment of the & connective). It is polarized and is either a focus, i.e. an address, together with a finite set of biases called ramification, or a daimon (necessarily positive) noted* ✠. *A* **chronicle** *has as* base *a set of addresses noted* $\Upsilon \vdash \Lambda$, *where* Υ *is either empty (positive base) or a unique address (negative base),* Λ *is a finite set of addresses. A chronicle is given as a sequence of actions with distinct focuses:*

- *alternating positive and negative actions, when the first one should be negative when the base is negative,*
- *justified: if present, a daimon should end the chronicle. A focus should be an address built by one of the previous actions (w.r.t. the sequence) or present in the base. The focus of a negative action should be either in the base (and the action is the first in the sequence) or generated by the action just before in the sequence.*

A **design** *on a basis* $\Upsilon \vdash \Gamma$ *is a set of chronicles such that*

- *the set is prefix closed, with branches on positive actions,*
- *a splitting property is satisfied: having negative actions just after a branching with distinct foci imposes distinct foci in the rest of the sequence of actions,*
- *branches in the arborescence end with a positive action,*
- *the set is non-empty in case the base is positive.*

In the following, designs are presented as arborescences (see Fig. 2).

Interaction, i.e. cut elimination (see Fig. 3), is defined between particular nets of designs [5]. First of all, addresses in bases should be distinct or present once in a left part of a base and once in a right part of another base, hence define a *cut*. The net of designs should be acyclic and connex w.r.t. the graph of bases and cuts. Therefore, one main design is distinguished either because it has a left part not part of a cut or it has no left part:

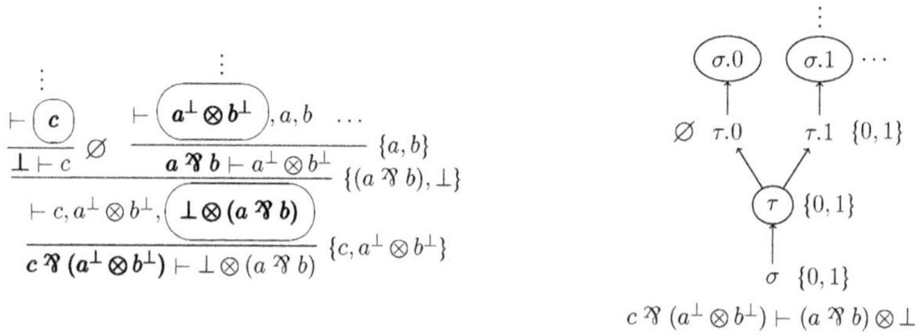

In the proof on the left, focuses are in bold font, rules are given w.r.t. the decomposition of the focus. Addresses replace formulas in the design given in figure on the right where the base is mentioned below. Addresses and formulas are circled whenever positive. Note that negative formulas are put on the left of the thesis sign (even they behave as if they were put on the right) to emphasize the fact that there cannot be more than one negative formula.

Fig. 2. From a (Linear Logic) proof to a (Ludics) design

Definition 2. *Let $\mathfrak{D}$ be the main design of a net of designs π, with first action (σ, I) or daimon,*

- *If the action is a daimon, the result is the design reduced to the daimon.*
- *If the focus of $\mathfrak{D}$ is not part of a cut, the result is obtained by applying this rule to the result of interaction to subdesigns (commutation step).*
- *Otherwise $\mathfrak{D}$ has no left part and its focus σ is part of a cut with another design with last rule $(\sigma, \mathcal{N})$ (aggregating ramifications of actions on the same focus σ):*
  - *If $I \notin \mathcal{N}$, then interaction fails.*
  - *Otherwise, interaction follows with the connected part of subdesigns obtained applying the part I of $\mathcal{N}$.*

Following this definition, either interaction fails, or it does not end ($\pi \Uparrow$), or it results in a design that reduces to ✠ when the net is closed (i.e. no commutation steps occur during the interaction). Two designs are orthogonal when their interaction reduces exactly to ✠. Finally *behaviours* denote formulas, where a behaviour is a set of designs equal to its biorthogonal. Referring to the Curry-Howard isomorphism, behaviours may be used as types for web processes.

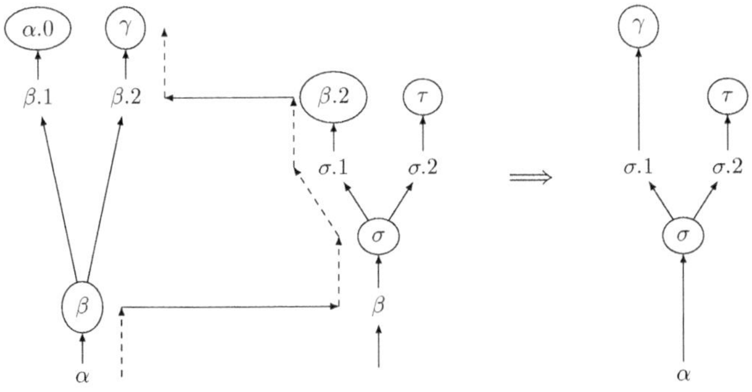

The figure presents a cut elimination between two designs (on the left), and its result (on the right). Commutations steps are marked as dashed arrows whereas other steps are plain lines. For sake of clarity, we omit ramifications beside nodes.

Fig. 3. Cut elimination between two designs

3.2 Interaction and Web Site

Our motto may be: A web site or web service is an alternating of actions/reactions between two tree-like structures modelling client and server sequences of operations. The following table summarizes the correspondances we put forward between Ludics objects, proof theory and web programming:

Ludics:	from *designs*	to *behaviours*	w.r.t. *interaction*
Proof theory:	to *proofs*	from *formulas*	w.r.t. *rules*
Web programming:	from *sequences of operations*	to *contracts*	w.r.t. *requests*

The model of web sites is not equivalent to web services where the client is a program as the server is. Agreeing the contract terms between a web service and a client means the computation, say the interaction, between the client and the server goes without execution errors. Obviously, one can abstract requests and arguments as function calls giving to them types as usual. Having in mind the fact that the process *should* continue (and not reduces to the computation of some result), one can type server and client with sequences alternating between requesting a call and proposing a set of available continuations. These types may be computed in advance, hence the compliance between a server and a client may be checked by simulating the interaction between server and client types, i.e. reducing the cut in terms of Ludics: either it fails and the interaction cannot take place, or the computation can begin (divergence cannot occur as types are supposed to be finite). Going back to web sites, a dynamic type checking should be done as nothing prevents the user from writing its own URL request to the server. Even though, requests afforded by the server are presented to the user for her to choose one following among what is presented: a web page is nothing else but available continuations, maybe embellished by html (static) data.

In the following, site addresses are our Ludics loci whereas requests are interaction calls. Site addresses are noted $u, v, w, \ldots$ on the server side and $a, b, c, \ldots$ on the user side. The syntax of requests is identical for server and user: $\beta@v\#c$ where β is a tag

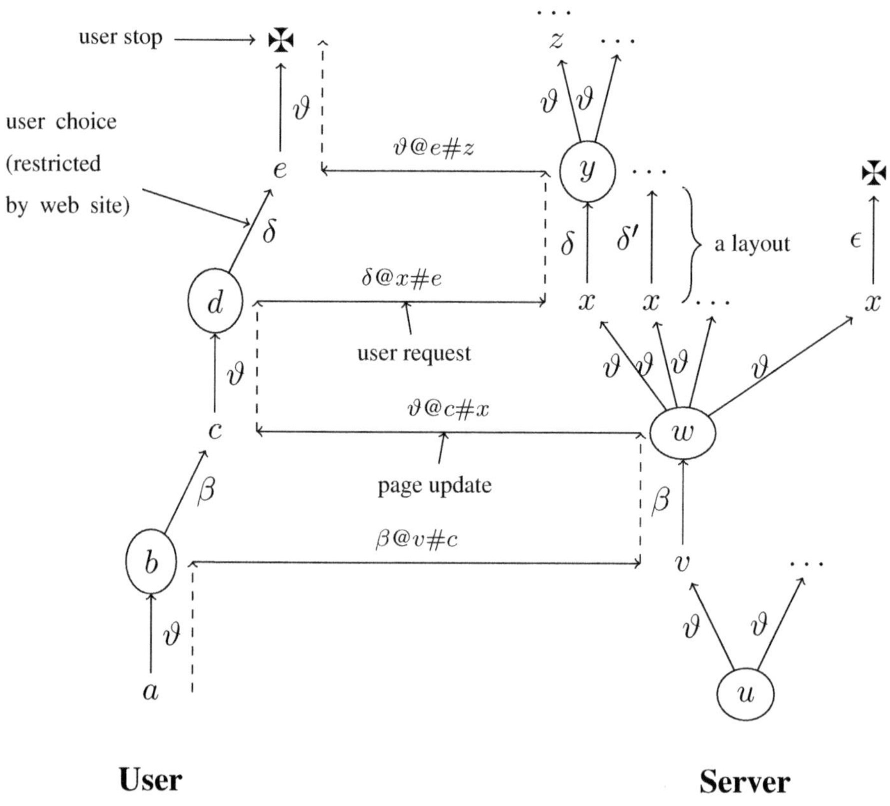

Fig. 4. A simple Web site dialogue between a user and a server

(a request name, i.e. a sub-locus index), v is the address of the receiver of the request, and c is the address waiting for the result. Requests may also contain a *body*, viewed as parameter values.[4] Expliciting addresses c and v in the request is necessary to cope with the distributed situation. In fact, memory spaces being distinct, duality between addresses (say equality except polarity in terms of Ludics) has to be made explicit: sending a request $\beta @ v \# c$ contains information that address c is the sub-address obtained after using tag β, hence equivalent to sub-address β of v. In web sites, the situation being asymmetric between client and server, concrete requests are particular:

- Sent by the server, the only tag at disposal to build a request is ϑ that requires the user browser to *update* the user display by data contained in the body of the request (but see below for other possibilities). Such a body contains tags at disposal for the next user request, i.e. links in the page to be displayed.
- Sent by the user, the tag used in the request corresponds to the user choice, i.e. her strategy w.r.t. this dialogue. Obviously, it is preferable that this tag be in the set of tags previously given by the server, but nothing prevents the user from another choice.

In terms of Ludics, actions defining user and server designs may then be completely specified:

- (user side) a positive action is either a daimon or a focus together with the link 'chosen' by the user as (part of) ramification. The ramification may not be a singleton in case values have to be sent to the user. Negative actions have all $\{\vartheta\}$ as ramification if *update* is the only action allowed by the browser to the server.
- (server side) Daimon is used to model a wrong page request. Otherwise the ramification of a positive action contains ϑ (call for an update of the browser). In such a design, negative actions occurring after some positive action have all the same focus (when the ramification of a positive action is a singleton). They form a sum of ramifications, each a singleton if complementary values are not requested, and as such define the *layout* of the page to be displayed by the user browser (with texts, ...).

A graphical example of typical interaction between a user and a web site is given Fig. 4. Let us begin when request $\vartheta @ c \# x$ is sent. This is an answer of the server interpreted as an update request toward the browser of the user. c is the state of the browser waiting for this update and d is its state after this update. The layout is concretely passed as a value and exposed to the user by the browser. It contains requests available from the server and presented as hyperlinks. An action at a positive user node may be either a daimon or with a ramification leading to negative nodes. The daimon is the situation occurring after negative node e: the user quits the dialogue. The other case occurs at nodes b and d in the user design. One can distinguish two main situations: (i) The user chooses a link accepted by the server (say δ at stage d); if coherence is checked between links present in the layout and server functionalities available at that moment, the link is in the layout. Interaction proceeds: a result is built by the server and sent next to the user. (ii) The user constructs a request that does not correspond to server possibilities.

[4] Parameter values may also be retrieved by means of additional requests, cf. subsection 3.3.

Then the interaction fails if that tag is not present on the server side, or the dialogue stops if the server anticipated wrong user requests and prepared a specific end (see the daimon following tag ϵ after negative server node x). In the two cases, arguments may be sent (cf. GET and POST html operations) or put at disposal for the server (as cookies memorized on the user side): in this last case, the ramification is no more a singleton and values should be retrieved by the server later during the interaction by additional requests.

An update request is in fact not the only kind of request the server may send: the browser may be asked to refresh the display, to open a new window or a new tab in the browser, ... For these cases to be taken into account, on the user side after some positive node, several negative nodes may be given with the same focus and distinct ramifications: each such node answers a server request interpreted by the browser.

Back, reload, and in general usage of a browser history, correspond to the following situations: the user being at a positive node (say s) sends a request already used (say $\beta@v\#e$ for a reload). In web situations, a page update follows and the user may continue with the link already chosen in the past or another link (or stop of course). Such a reuse of old addresses is not linear and thus impossible as such in the Ludics paradigm. The only safe possibility (w.r.t. linear interaction) is to consider that the current interaction is suspended, a new one is launched with the same steps as the previous one. In that case, a renaming of addresses occurs to guarantee old and new server and client designs differ.[5] Note that renaming of addresses is a safe operation on designs [10]. Remark finally that it corresponds to the functional implementations we discussed in Section 2. It is also a basic ingredient of our FICX language. Session identifiers serve also to implement such a renaming (they are included in the request) when the browser generates a fresh one. However most of current implementations rely on cookies or session identifiers that do not allow the renaming to guarantee the correctness of interaction.

3.3 From Ludics Model to FICX Programming Language

K. Terui proposed in [10] a reformulation of Ludics more suitable from a computational point of view. *c-designs* are built as Girard's designs except that (i) variables may occur as generic addresses (possibly in an infinite number), (ii) positive nodes may be explicit *internal* cuts or serve to model divergence of interaction, (iii) c-designs may not be linear. The syntax of c-designs is such that they may be more easily used for programming purposes, and a grammar allows for a finite specification of them by means of *generators*. K. Terui uses this model for characterizing word languages where the acceptance relation between a word and a grammar is expressed w.r.t. orthogonality between c-designs. c-designs is the core of the type language given for *Xobjects* in the next section. Xobjects should be interpreted as the computational object able to answer user requests, hence has as type a negative (server) node in terms of designs. Besides this aspect, Xobjects include also xml or html data to be published in a browser, and they are parameterized in such a way that modular programming and frame programming are facilitated.

[5] In fact, one may keep the same addresses until the requested address.

Definition 3. *Let T be a set of names ($\alpha, \beta, \gamma, \ldots$, denoting page tags) and a function ar giving for each tag name in T an integer, let $\mathcal{V}$ be a set of variables ($x, y, \ldots$). Addresses are denoted as before $a, b, \ldots, u, v, \ldots$ and are used as references for positive and negative c-designs. The sets of positive and negative c-designs are inductively defined by:*

$P ::= \maltese$	`daemon`
$\mid \Omega$	`divergence`
$\mid N_0 \mid \gamma < \overrightarrow{N_\gamma} >$	`internal cut if` N_0 `is not a variable`
$N ::= x$	`variable`
$\mid \sum \gamma(\overrightarrow{x_\gamma}) \# P_\gamma$	`sum of functionalities (e.g. links or browser capabilities) afforded at some stage possibly with required arguments`

In the previous definition, $\overrightarrow{N_\gamma}$ (resp. $\overrightarrow{x_\gamma}$) is a finite sequence $N_1, \ldots, N_n$ (resp. $x_1, \ldots x_n$) such that $ar(\gamma) = n$. Finally, the sum is over a set of tag names. If we go back to example given Fig. 4, the web server may be defined in the following way ('[' and ']' surround terms, node names as mentioned in Fig. 4are given as upperscript to '[' such that $[^u\, blabla]$ means that u is a pointer referring to term $blabla$, u is passed from/to server and client instead of its content):

$$
\begin{array}{l}
[^u\ x \mid \vartheta < [^v\ \beta()\#[^w\ y \mid \vartheta < [^x\ \delta()\#[^y\ z \mid \vartheta < [^z\ \ldots] >] \\
\qquad\qquad\qquad\qquad\qquad\qquad\qquad\quad +\delta'()\#\ldots \\
\qquad\qquad\qquad\qquad\qquad\qquad\qquad\quad +\ldots \\
\qquad\qquad\qquad\qquad\qquad\qquad\qquad\quad +\epsilon()\#\maltese\quad] >]] >]
\end{array}
$$

Interaction between a negative c-design $\sum \gamma(\overrightarrow{x_\gamma})\#P_\gamma$ and a positive c-design $x \mid \delta < \overrightarrow{N_\delta} >$ is equivalent to computing the reduction of the positive c-design $\sum \gamma(\overrightarrow{x_\gamma})\#P_\gamma \mid \delta < \overrightarrow{N_\delta} >$:

Definition 4. *The reduction rule is given by ($i \in [1, ar(\gamma)]$):*

$$
\sum \gamma(\overrightarrow{x_\gamma})\#P_\gamma \mid \gamma < \overrightarrow{N_\gamma} > \quad \longrightarrow \quad P_\gamma[N_i/x_i]
$$

Normalization succeeds (i.e. interaction succeeds) if $P \longrightarrow^* Q$ and Q is neither a cut nor Ω, where $\longrightarrow^*$ is the transitive closure of $\longrightarrow$.

Design generators are defined by Terui [10] to allow recursive c-designs *definitions*. Fresh addresses are created each time a definition is used. This is close to the way variables are used during resolution in logic programming. Hence a design generator is nothing else but a definition of a c-design, maybe referring other design generators. This is particularly useful in concrete (web sites) situations where links may lead to previously defined pages (however not with the same address when interaction goes on). We do not develop their formal aspects even if their use is essential when designing a concrete programming language. Note that this appears as a convenient way to deal with (maybe infinite) designs but essential logical properties remain.

Program	
P ::= ϵ	*empty program*
$\mid S\ P \mid d\ P$	*Xstructure or definition followed by a program*
Xstructure	
S ::= `xstruc` d `begin` $w = e$	*where d is a definition, w an identifier, e an expression*
Xobject	
e ::= `xobject`$\langle Y_1, \ldots, Y_n \rangle$	*Xobject definition with abstract paths parameters*
e ▶ sr	*Xdata* ▶ *reactions*
`xend`	
$\mid e_1[Y \mapsto e_2]$	*parameter assignment*
$\mid \tau(e_2)@e_1$	*request evaluation*
$\mid Y$	*Abstract path name Y*
Reactions	
sr ::= ϵ $\quad$ r ::= $\tau_p(p) \Rightarrow e$	*reaction conditioned by trigger and parameter patterns*
$\mid r\ sr$	
Triggers	
τ ::= $Y.C$	*abstract path followed by a tag*

Fig. 5. Grammar of FICX

4 FICX: A Web Functional Langage

The language FICX, *Functional Interactive and Compositional XML*, is devoted to Web programming [3]. It is the result of a joint work with P. Coupey and J.-V. Loddo, done independently from Ludics. However, the type system of FICX fits well into a Ludics interpretation of interaction. FICX uses a functional programming language, currently OCaml, as a core language for functions, definitions, ... This core is extended by means of an *Xobject* data structure that integrates an extended XML structure called *Xdata* to publish data and hyperlinks, and a functional part called *reaction* intended to answer requests built from hyperlinks. Moreover Xobjects may be parameterized by *abstract paths* defined below. Finally, an *Xstructure* is a specific top-level definition for specifying interactive data and functionalities to be used by clients. The grammar of the language FICX, specific to our aim, is given in Fig. 5. We use the following notations: e is an expression and p is a pattern, a, A, C, x, y are identifiers, τ is a trigger[6], Y, Z are abstract paths, finally r states for a reaction. Abstract paths are defined according to the following grammar, where y is an identifier, `parent`, `root` and `self` are keywords: Y ::= `parent` | `root` | `self` | y | Y.Y Let us define the following rewriting: y.`parent` $\rightarrow \epsilon$, Y.`root` $\rightarrow$ `root` , Y.`self` $\rightarrow$ `self` , it gives rise on abstract paths (and abstract path patterns or path types in the same way) to two kinds of normal forms: `root`.Y and `self`.Y with `self` and `root` not in Y. The intended meaning is that a set of such abstract paths should partially define two rooted trees, one with abstract root `root` and another 'centered' on `self` where `self.parent`. . . .`parent` should represent a path 'up' to some concrete root. Abusively, `self` may be omitted in the following from abstract paths writings.

An *Xobject* is structured in two parts: an *Xdata* structure together with *reactions*, and is parameterized by means of abstract paths. Parameterization is a convenient way to refer to yet unknown Xobjects while parameter assignment merges partial trees of

[6] A trigger is a tag name maybe preceded by an abstract path.

```
xstruc
link =   xobject <root.H1> (x:string)
         root.H1.T<a>[x]
     ▶ xend;
message = xobject <> (msg:string)
         <h1 align="center">[msg]
     ▶ xend;
phandler = xobject <M1> (k:int)
         M1 <p>["Visits for this session (cs): " k]
     ▶
         T(<a>[x]) ⇒ (let y=(if (x = "Hello") then "Salut" else "Hello") in
                         (phandler (k + 1))[M1↦(message y)])
    xend;
home = xobject <L1, L2, H1>
         <html>[ <head>[ <title>["Welcome"]] <body>[ H1 <br> L1 <br> L2]]
     ▶ xend;

m1 = (message "Hello");
h1 = (phandler 0)[M1 ↦ m1];
l1 = (link "Increment cs and reload with Hello");
l2 = (link "Increment cs and reload with Salut");
o = home[L1 ↦ l1][L2 ↦ l2][H1 ↦ h1];

begin website = o
```

Fig. 6. An example of a website definition in FICX

components. Abstract paths that are used in an Xobject body are declared in the header: in Fig. 6, Xobject `home` expects three subcomponents `L1`,`L2`,`H1`. Note the reference to `root.H1` in the definition of `link`: this Xobject is expected to be in a tree whose Xobject root has at least a subcomponent for label `H1`. `website` is declared as an entry point. The tree of components is rooted at `o`, that has two links `l1` and `l2` and one phandler `h1` as subcomponents, `h1` having a child `m1`. Two triggers are declared, posted in `l1` and `l2`, authorizing interactive requests of the form `H1.T(x)@website`.

The operational semantics corresponding to an Xobject declaration is straightforwardly a closure (as for functions) and assignment of abstract path parameters is similar to the standard treatment of handlers in functional programming. The language for the Xdata part extends XML: an abstract path may be used in place of an XML node and each node in an XML structure may be labelled by a trigger τ. A trigger τ has the general form $Y.C$ where C is an *interaction tag* (tag for short) and the abstract path Y is the path to the *abstract last possible receiver*. Setting down a trigger means that a functionality should be available, as a GET in HTML or the description of a service. In case of Xstructures, a built-in function `get_Xdata` is available that extracts the Xdatum to build a (standard) XML structure that can be sent interactively. When encountering an abstract path in place of an XML node, the function is recursively called on the value at the abstract path. In case of a trigger, a request is prepared that includes the address of the value of `root` (called the *initial concrete receiver*), the abstract path from it to the last possible receiver, the interaction tag and an XML structure (the parameter of the request). Note that the abstract path to the last possible receiver is now given from `root`, then may be different from the abstract path set up in the Xdata. Such an interactive request is at first received (and executed) by the initial concrete receiver. In fact a request is a first-class citizen that has the general form $\tau(e_2)@e_1$: e_1 is the concrete receiver of the request, $\tau(e_2)$ gives the trigger and the parameter value for the request.

The operational semantics of a request may be given informally in the following way: if the concrete receiver has an adequate reaction (the reaction matches trigger and parameter of the request), the reaction is evaluated ; otherwise the request is delegated following the path to the last possible receiver until some Xobject has an adequate reaction. It is the type system that is responsible for checking that there cannot be run-time errors. Note that an Xobject value is rebuilt when a delegation occurs. Going back to Fig. 6, a tag `T` is set down in the Xdata part of `link`: **`root`**`.H1.T<a>[x]` states that the tag `T` is an anchor. Requests available in this case may be for example `T("Hello")@h1` or `H1.T("Hello")@website`. This last request is the only one that can be used interactively. As `website` has no appropriate reaction, the request is delegated to `h1`, value of `H1` as it is given in `website`. The reaction part of this phandler contains a reaction that is fired with result the Xobject `(phandler 1)[M1↦(message "Salut")]`. The final value sent back to the requester is `home[L1↦l1][L2↦l2][H1↦(phandler 1)[M1↦(message "Salut")]]`.

The type system associated is closely related to what has been set up in previous sections (see Fig. 7). An Xobject type consists of a type for its Xdata part together with a type for its reactions. The type for its set of reactions is a sum over the available reactions that should be declared as hyperlinks in the Xdata part. Each reaction is indexed by a tag and may require arguments. The result of a reaction is an Xobject. Requests *inside* the programming language are typed as internal cuts. The only extension done to the language of c-designs concerns the need to type the tree-like structure of Xobjects. The complete type system may be found in [3]. It is not too difficult to prove a safety theorem stating that well-typed expressions are evaluable, i.e. there cannot be evaluation errors (provided for the functional language part an operational semantics safe with respect to a standard typing):

Xdata	
$t^s \ ::= [\tau]{<}a\ l_t{>}[r_t]$	*Xdata tree*
$\quad \mid\ Y$	*abstract path type name*
Xobjects	
$t^o \ ::= \mathbb{T} \mid Y$	*an Xobject is an abstract tree or a path type*
	where an abstract tree is a function
	*from paths to **T**node values*
$vn ::= t^s \blacktriangleright \rho$	*type of a **T**node value*
$\rho \ ::= \epsilon$	*reactions are a sequence of Xobject types*
$\quad \mid\ \tau_{p1} \rightarrow t^s_{p1}\ [\rightarrow t^o_1]; \ldots ;\tau_{pn} \rightarrow t^s_{pn}\ [\rightarrow t^o_n]$	*for a pattern type for trigger type*

Fig. 7. Type language of FICX (except XML and functional type language)

Theorem 1. *Let* e *be an expression of the language, if* $\vdash e : t$ *is provable, i.e. the expression is typable, then there exist a value* v *and a handler environment* $\mathcal{H}'$ *such that* $\vdash e \Downarrow v, \mathcal{H}'$ *is provable, i.e. the expression may be evaluated without errors.*

5 Towards Web Services

Web Services have been largely developed since the first examples programmed in 1997 with the framework Jini. It was originally designed to overcome problems due to security and code mobility in distributed situations. A *web service* is currently given as a name, i.e. a URL address, a program, a contract, maybe with a list of semantic attributes. Semantic attributes are intended to abstract from concrete names in favor of a characterization of the kind of service that is provided, e.g. 'train timetable', 'notebook furnisher', ... On the contrary, a contract specifies the model of interaction the web service can have with a client. A *client* of a web service is given as a program together with a co-contract, that is a model of interaction in the point of view of such a client.

The operational steps may be summarized as follows: (i) The server providing the web service *publishes* its name, contract and possibly the list of attributes to a dedicated publisher server. (ii) When a client program searches for a specific web service, it sends a request to the publisher server to retrieve the list of web services satisfying a given name or a given list of attributes. In particular, it gets a list of contracts. (iii) The client checks the compatibility between contracts and its own co-contract, and chooses a web service satisfying its requirements. (iv) Interaction begins between the client and the selected web service until the end. We do not develop steps (i) and (ii) in this paper and only turn our attention towards the content of contract and co-contract. Contract and co-contract differ in that it is the client responsability to begin and quit a conversation. Having that in mind, we can forget about these differences and only speak of contracts as soon as a language for contracts integrates the two point of views. Even if contracts and co-contracts are concretely abstract descriptions of programs, we should remark that the semantics of current standards is far from being completely formal. Various contract description languages exist, WSDL, WSCL, BPEL among others. WSDL, *Web Services Description Language*, is still widely used even if extremely limited as the specification of an interaction is limited to the description of a single exchange between the client and the server: at most one output of values and/or one input of results. A contract is written as an XML document. Available functionalities are named *operations*, each of them precises if it is a one-way or a two-way method (the pattern gives its semantics). It may include options or treatments of exceptions (called robustness) as well as argument types. On the contrary, WSCL (*Web Service Conversation Language*) and BPEL (*Business Protocol Evaluation Language*) are sufficiently rich to support abstract description of the full interactive behaviour of a program (case structures, loops, ...). BPEL focuses on combining web services to better specify their *orchestration*. One may find at `http://docs.oasis-open.org/wsbpel/2.0/OS/wsbpel-v2.0-OS.html`[7] examples of BPEL processes. Documents necessary for such web services include one document for a semantic description, another one for the low-level protocols to be used, ... and a document describing the process itself. This last document begins with a definition of types in use in the document, followed by the names of variables and *partners* (say the client, the server, ...). Finally a part describes the behavior of the interaction as a program schema.

Carpineti et al. [2] define a process language devoted to WSDL web services in a fragment of CCS. Contracts are abstract definitions of such process descriptions. An example is given Fig. 8. Contracts are either void (end of a computation), or begin with a named request (e.g. waiting for an input for tag A) or result (e.g. output of the value of tag $\overline{\mathrm{A}}$), or are given as choices among different contracts. Choices may be external

[7] XML documents are generally verbose. It is worse with web service descriptions so we omit to include even a simple example in this paper.

$$
\begin{array}{llll}
\texttt{Login.(} & \overline{\textsc{InvalidLogin}}\ .\ \overline{\textsc{End}} & & \\
& \oplus\ \overline{\textsc{ValidLogin}}\ .\ \texttt{Query}\ .\ \overline{\textsc{Catalog}}\ . & & \\
& \quad\texttt{(} & & \\
& \quad\quad \texttt{Logout}\ .\ \overline{\textsc{End}} & & \\
& \quad\quad +\ \texttt{Purchase}\ .\texttt{(} & \overline{\textsc{Accepted}}\ .\ \overline{\textsc{End}} & \\
& & \oplus\ \overline{\textsc{InvalidPayment}}\ .\ \overline{\textsc{End}} & \\
& & \oplus\ \overline{\textsc{OutOfStock}}\ .\ \overline{\textsc{End}} & \\
\texttt{)} & \quad\texttt{)} & \texttt{)} &
\end{array}
$$

Fig. 8. A contract for a simplified web purchase service [2]

(connective $+$) or *internal* (connective $\oplus$). External choice means a partner choice, whereas internal choice is done locally. Carpineti et al. defines a compliance relation compatible with the operational semantics of the CCS fragment they use for modelling web service processes: a contract c is compliant with a contract d if operations with c are possible with d, whence the same operational compatibility between processes. Their specification of a web service contract, therefore of a compliance relation, has a few drawbacks among which the fact that it is not transitive. As authors notice "The lack of transitivity has a non negligible impact on the use the relation. For instance, while it is possible to replace a given service with a new service whose subcontract is greater than the original service's contract, it is not possible to renew this operation without taking into account the original contract."

We consider the problem may be overcome by rebuilding a contract language in the same spirit we do for web sites. In particular, this supposes that polarity has to be strictly taken into account contrarily to Carpineti et al.' contract language (see example Fig. 8 where actions of identical polarities follow). The $+$ connective is straightforwardly the sum as given in subsection 3.3. The use of $\oplus$ suggests to view a contract as a set of designs (distinguishing the various interactions w.r.t. local choices). Beginning with designs, contracts are then biorthogonal of sets of designs, hence transitivity follows as a general property of behaviours.

This line of research may help defining a fully distributed language built over the concept of interaction. Moreover, it allows for an identical view of web sites and web services. One may try also to extend such a point of view to other program-to-program and program-to-user cases when non textual information is present, e.g. the window system for computers or even distributed ontologies in Web 2.0. Let us summarize the benefits of interpreting computations in web site or web service domains as (c-Ludics) dialogues: (i) the overall framework has nice logical properties, hence it induces safe and natural type systems, (ii) a large set of combined dialogues/computations may be modelled giving a clean meaning to distributed computational situations, (iii) behaviours are closures of sets of designs and denote formulas, hence induce a safe notion of contract, (iv) interaction as well as expression calls (internal to the language) are represented in a coherent way with c-designs as proposed by K. Terui. Nevertheless, it is clear that the restriction of Ludics to linearity is a limit in some cases: the reusability of memory objects may well be clearly safe but difficult to grasp without exponentials.

References

[1] Andreoli, J.-M.: Logic programming with focusing proofs in linear logic. J. Log. Comput. 2, 297–347 (1992)
[2] Carpineti, S., Castagna, G., Laneve, C., Padovani, L.: A formal account of contracts for web services. In: Bravetti, M., Núñez, M., Zavattaro, G. (eds.) WS-FM 2006. LNCS, vol. 4184, pp. 148–162. Springer, Heidelberg (2006)
[3] Coupey, P., Fouqueré, C., Loddo, J.-V.: Tree components programming: An application to xml. In: Jones, C.B., Liu, Z., Woodcock, J. (eds.) ICTAC 2007. LNCS, vol. 4711, pp. 139–153. Springer, Heidelberg (2007)
[4] Girard, J.-Y.: Linear logic. Theor. Comput. Sci. 50, 1–102 (1987)
[5] Girard, J.-Y.: Locus solum: From the rules of logic to the logic of rules. Mathematical Structures in Computer Science 11, 301–506 (2001)
[6] Girard, J.-Y.: Le Point Aveugle: vers l'imperfection. Visions des Sciences 2, Hermann (2007)
[7] Hughes, J.: Generalising monads to arrows. Sci. Comput. Program. 37, 67–111 (2000)
[8] Queinnec, C.: The influence of browsers on evaluators or, continuations to program web servers. In: ICFP, pp. 23–33 (2000)
[9] Queinnec, C.: Continuations and web servers. Higher-Order and Symbolic Computation 17, 277–295 (2004)
[10] Terui, K.: Computational ludics. Theoretical Computer Science (2008) (to appear)
[11] Thiemann, P.: Wash/cgi: Server-side web scripting with sessions and typed, compositional forms. In: Krishnamurthi, S., Ramakrishnan, C.R. (eds.) PADL 2002. LNCS, vol. 2257, pp. 192–208. Springer, Heidelberg (2002)
[12] Thiemann, P.: Wash server pages. In: Hagiya, M., Wadler, P. (eds.) FLOPS 2006. LNCS, vol. 3945, pp. 277–293. Springer, Heidelberg (2006)
[13] Wadler, P.: Comprehending monads. In: LISP and Functional Programming, pp. 61–78 (1990)

On the Meaning of Focalization

Michele Basaldella[1,⋆], Alexis Saurin[2,⋆⋆], and Kazushige Terui[1,⋆]

[1] RIMS, Kyoto University
mbasalde@kurims.kyoto-u.ac.jp, terui@kurims.kyoto-u.ac.jp
[2] Università degli Studi di Torino
alexis.saurin@pps.jussieu.fr

Abstract. In this paper, we use Girard's ludics to analyze focalization, a fundamental property of the proof theory of linear logic. In particular, we show how focalization can be realized interactively thanks to suitable section-retraction pairs between semantical types.

1 Introduction

Focalization is a deep outcome of linear logic proof theory, putting to the foreground the role of polarity in logic. It resulted in important advances in various fields ranging from proof-search (the original motivation for Andreoli's study [1] of focalization) and the ability to define hypersequentialized calculi (via synthetic connectives) [6, 7] to game semantical analyses of logic.

In particular, focalization deeply influenced Girard's ludics [8] which is a pre-logical framework which aims to analyze various logical and computational phenomena at a foundational level. For instance, the concluding results of Locus Solum are a full completeness theorem with respect to focalized multiplicative–additive linear logic **MALL**.

Another characteristics of ludics is that types are built from untyped proofs (called *designs*). More specifically, types (called *behaviours*) are sets of designs closed under a certain closure operation. This view of types as sets of proofs opens a new possibility to discuss focalization and other properties of proofs *at the level of types*.

The purpose of this abstract is to show that ludics is suitable for analyzing focalization and that this interactive analysis of focalization is fruitful. In particular, our study of focalization in ludics was primarily motivated by the concluding remarks of the third author's paper on computational ludics [12] where focalization on *data designs* was conjectured to correspond to the tape compression theorem of Turing machines.

Still, for the very reason that ludics abstracts over focalization (being built on hypersequentialized calculi) it is not clear whether an analysis of focalization can (or shall) be pursued in ludics: an obstacle is, however, that ludics is already fully focalized, so that there seems to be no room to discuss and prove focalization

⋆ Supported by Grant-in-Aid for Scientific Research, JSPS.
⋆⋆ Supported by a Bourse Lavoisier.

Lecomte and S. Tronçon (Eds.): PRELUDE 2010, LNAI 6505, pp. 78–87, 2011.

internally. This can be settled by using a dummy shift operator. For instance, a compound formula $L \oplus (M \otimes N)$ of linear logic can be expressed in ludics in two ways; either as a flat behaviour $\oplus \otimes (L, M, N)$ built by a single synthetic connective $\oplus\otimes$ from three sub–behaviours L, M, N, or as a compound behaviour $L \oplus \uparrow (M \otimes N)$, which consists of three layers: $M \otimes N$ (positive), $\uparrow (M \otimes N)$ (negative), and $L \oplus \uparrow (M \otimes N)$ (positive).

Focalization can then be expressed as a mapping from the latter to the former behaviour. Hence we can deal with it *as if it were an algebraic law*, which may be compared with other logical isomorphisms such as associativity, distributivity, etc. To be precise, however, focalization is not an isomorphism but is an asymmetric relation. In this paper, we think of it as a *retraction* $L \oplus \uparrow (M \otimes N) \longrightarrow \oplus \otimes (L, M, N)$ which comes equipped with a section $\oplus\otimes(L, M, N) \longrightarrow L \oplus \uparrow (M \otimes N)$.

The aim of our current work is to promote this "algebraic" view of focalization in the setting of ludics. Furthermore, section-retraction pairs can be naturally lifted by applications of logical connectives (Theorem 3). Hence we also have focalization inside a compound behaviour (or inside a context). This would allow us to recover the original focalization theorem as a corollary to our "algebraic" focalization, though we leave it as future work.

2 Focalization in Linear Logic

Linear logic (**LL**) comes from a careful analysis of structural rules in sequent calculus resulting in a very structured proof theory, in particular regarding dualities. A fundamental outcome of those dualities is Andreoli's discovery [1] of focalization, providing the first analysis of polarities in **LL**. Andreoli's contribution lies mainly in the splitting of logical connectives in two groups – positive ($\otimes, \oplus, \mathbf{0}, \mathbf{1}, \exists, !$) and negative ($⅋, \&, \top, \bot, \forall, ?$) connectives – besides the at-the-time more traditional multiplicative–additive–exponential distinction.

The underlying meaning of polarization comes from proof-search motivations. The introduction rules for negative connectives $⅋, \&, \top, \bot, \forall$ are *reversible*: in the bottom–up reading, the rule is deterministic, i.e., there is no choice to make and provability of the conclusion implies provability of the premisses. On the other hand, the introduction rules for positive connectives involve choices: e.g., splitting the context in $\otimes$ rule, or choosing between $\oplus_L$ and $\oplus_R$ rules, resulting in the possibility to make erroneous choices during proof-search. Still, *positive connectives* satisfy a strong property called *focalization* [1]: let us consider a sequent $\vdash F_1, \ldots, F_n$ containing no negative formulas, then there is (at least) one formula F_i which can be used as a *focus* for the search by hereditarily selecting F_i and its positive subformulas as principal formulas up to the first negative subformulas.

This property induces the following strategy of proof–search called *focalization discipline*:

Sequent Γ contains a negative formula	Sequent Γ contains no negative formula
choose any negative formula (e.g. the leftmost one) and decompose it using the only possible negative rule	choose some positive formula and decompose it (and its subformulas) hereditarily until we get to atoms or negative subformulas

A sequent calculus proof is called *focussing* if it respects the focalization discipline. It is proven in [1] that if a sequent is provable, then it is provable with a focussing proof: the focalization discipline is therefore a complete proof-search strategy. Other approaches to focalization consider proof transformation techniques [10, 11].

A very important consequence of focalization is the possibility to consider *synthetic connectives* [7, 4]: a synthetic connective is a maximal cluster of connectives of the same polarity. They are built modulo commutativity and associativity of binary connectives and some syntactical isomorphism [9] of linear logic. For **MALL**, the underlying syntactical isomorphism in action is the *distributivity* of $\otimes$ with respect to $\oplus$, namely $(A \oplus B) \otimes C \cong (A \otimes C) \oplus (B \otimes C)$ and its dual.

3 Ludics in Three Pages

Syntax. We recall the term syntax for designs introduced in [12] which uses a process calculus notation inspired by the close relationship between ludics and linear π-calculus [5].

Designs are built over a given *signature* $\mathcal{A} = (A, \mathsf{ar})$, where A is a set of *names* $a, b, c, \ldots$ and $\mathsf{ar} : A \longrightarrow \mathbb{N}$ assigns an *arity* $\mathsf{ar}(a)$ to each name a. Let $\mathcal{V}$ be a countable set of variables $\mathcal{V} = \{x, y, z, \ldots\}$. Over a fixed signature $\mathcal{A}$, a (proper) *positive action* is $\overline{a}$ with $a \in A$, and a (proper) *negative action* is $a(x_1, \ldots, x_n)$ where $x_1, \ldots, x_n$ are distinct variables and $\mathsf{ar}(a) = n$. In the sequel, an expression of the form $a(\vec{x})$ always stands for a negative action.

The positive (resp. negative) *designs* P (resp. N) are coinductively generated by the following grammar (where $\mathsf{ar}(a) = n$ and $\vec{x} = x_1, \ldots, x_n$):

$$P ::= \Omega \text{ (divergence)} \;\big|\; \maltese \text{ (daimon)} \;\big|\; N_0 | \overline{a}\langle N_1, \ldots, N_n \rangle \text{ (application)},$$
$$N ::= x \text{ (variable)} \;\big|\; \sum a(\vec{x}).P_a \text{ (abstraction)},$$

where the formal sum $\sum a(\vec{x}).P_a$ is built from $|A|$-many components $\{a(\vec{x}).P_a\}_{a \in A}$. Designs may be considered as infinitary λ-terms with *named* applications and *superimposed* abstractions. $P, Q, \ldots$ (resp. $N, M, \ldots$, resp. $D, E, \ldots$) denote positive (resp. negative, resp. arbitrary) designs. Any subterm E of D is called a *subdesign* of D. Ω is used to encode partial sums: given a set $\alpha = \{a(\vec{x}), b(\vec{y}), \ldots\}$ of negative actions, we write $a(\vec{x}).P_a + b(\vec{y}).P_b + \cdots$ to denote the negative design $\sum_\alpha a(\vec{x}).R_a$, where $R_a = P_a$ if $a(\vec{x}) \in \alpha$, and $R_a = \Omega$ otherwise.

A design D may contain free and bound variables. An occurrence of subterm $a(\vec{x}).P_a$ *binds* the free-variables $\vec{x}$ in P_a. Variables which are not under the scope of the binder $a(\vec{x})$ are *free*. $\mathsf{fv}(D)$ denotes the set of free variables occurring in D. Designs are considered *up to α-equivalence*, that is up to renaming of bound variables (see [12] for further details).

A positive design which is neither Ω nor $\maltese$ is either of the form $(\sum a(\vec{x}).P_a) | \overline{a}\langle N_1, \ldots, N_n \rangle$ and called a *cut* or of the form $x | \overline{a}\langle N_1, \ldots, N_n \rangle$ and called a *head normal form*. The *head variable* x in the design above plays the same role as a pointer in a strategy does in Hyland-Ong games and an address (or locus) in Girard's ludics. On the other hand, a variable x occurring in a bracket (as in $N_0|\overline{a}\langle N_1, \ldots, N_{i-1}, x, N_{i+1}, \ldots, N_n\rangle$) does not correspond to a pointer nor address but rather to an identity axiom (initial sequent) in sequent calculus, and for this reason is called an *identity*.

A design D is said: *total*, if $D \neq \Omega$; *linear* (or *affine*), if for any subdesign of the form $N_0|\overline{a}\langle N_1, \ldots, N_n\rangle$, the sets $\mathsf{fv}(N_0), \ldots, \mathsf{fv}(N_n)$ are pairwise disjoint.

Normalization. The *reduction relation* $\longrightarrow$ is defined on positive designs as follows:

$$\left(\sum a(x_1, \ldots, x_n).P_a\right)|\overline{a}\langle N_1, \ldots, N_n\rangle \longrightarrow P_a[N_1/x_1, \ldots, N_n/x_n].$$

We denote by $\longrightarrow^*$ its transitive closure. Given two positive designs P, Q, we write $P \Downarrow Q$ if $P \longrightarrow^* Q$ and Q is neither a cut nor Ω. We write $P \Uparrow$ if there is no Q such that $P \Downarrow Q$.

The *normal form function* $[\![\ ]\!] : \mathcal{D} \longrightarrow \mathcal{D}$ is defined by corecursion as follows:

$$\begin{array}{lll|rl} [\![P]\!] & = \maltese & \text{if } P \Downarrow \maltese; & [\![\sum a(\vec{x}).P_a]\!] & = \sum a(\vec{x}).[\![P_a]\!]; \\ & = x|\overline{a}\langle[\![\vec{N}]\!]\rangle & \text{if } P \Downarrow x|\overline{a}\langle\vec{N}\rangle; & [\![x]\!] & = x. \\ & = \Omega & \text{if } P \Uparrow; & & \end{array}$$

Normalization is *associative*:

$$[\![D[N_1/x_1, \ldots, N_n/x_n]]\!] = [\![[\![D]\!][[\![N_1]\!]/x_1, \ldots, [\![N_n]\!]/x_n]]\!].$$

Orthogonality. In the rest of this work, we restrict ourselves to the special subclass of *total*, *cut-free*, *linear* and *identity-free* designs (corresponding to [8]).

A positive design P is *closed* if $\mathsf{fv}(P) = \emptyset$, *atomic* if $\mathsf{fv}(P) \subseteq \{x_0\}$ for a certain fixed variable x_0. This variable plays the same role as the empty address $\langle\rangle$ in [8].

A negative design N is *atomic* if $\mathsf{fv}(N) = \emptyset$. Two atomic designs P, N of opposite polarities are said *orthogonal* (written $P \bot N$) when $[\![P[N/x_0]]\!] = \maltese$. If $\mathbf{X}$ is a set of atomic designs of the same polarity, then its *orthogonal set* is defined by $\mathbf{X}^{\bot} := \{E : \forall D \in \mathbf{X},\ D \bot E\}$. Although possible, we do not define orthogonality for non-atomic designs. Accordingly, we only consider *atomic* behaviours which consist of atomic designs.

An (atomic) *behaviour* $\mathbf{X}$ is a set of atomic designs of the same polarity such that $\mathbf{X}^{\bot\bot} = \mathbf{X}$. A behaviour is positive or negative according to the polarity of its designs. We denote positive behaviours by $\mathbf{P}, \mathbf{Q}, \mathbf{R}, \ldots$ and negative behaviours by $\mathbf{N}, \mathbf{M}, \mathbf{K} \ldots$.

There are the least and the greatest behaviours among all positive (resp. negative) behaviours with respect to set inclusion (with $\maltese^- = \sum a(\vec{x}).\maltese$):

$$\mathbf{0}^+ := \{\maltese\}, \qquad \mathbf{0}^- := \{\maltese^-\}, \qquad \top^+ := \mathbf{0}^{-\bot}, \qquad \top^- := \mathbf{0}^{+\bot}.$$

A *positive sequent* $\mathbf{\Gamma}$ is a set of the form $x_1 : \mathbf{P}_1, \ldots, x_n : \mathbf{P}_n$, where $x_1, \ldots, x_n$ are distinct variables and $\mathbf{P}_1, \ldots, \mathbf{P}_n$ are (atomic) positive behaviours. We denote by $\mathsf{fv}(\mathbf{\Gamma})$ the set $\{x_1, \ldots, x_n\}$. A *negative sequent* $\mathbf{\Gamma}, \mathbf{N}$ is a positive sequent

$\mathbf{\Gamma}$ enriched with an (atomic) negative behaviour $\mathbf{N}$, to which no variable is associated.

A positive design P *belongs* to a positive sequent $\mathbf{\Gamma}$ (notation: $P \models \mathbf{\Gamma}$) if $\mathsf{fv}(P) \subseteq \{x_1, \ldots, x_n\}$ and $[\![P[N_1/x_1, \ldots, N_n/x_n]]\!] = \maltese$ for any $N_1 \in \mathbf{P}_1^{\perp}, \ldots, N_n \in \mathbf{P}_n^{\perp}$.

Analogously, we write $N \models \mathbf{\Gamma}, \mathbf{N}$ whenever $\mathsf{fv}(N) \subseteq \{x_1, \ldots, x_n\}$ and $[\![P[N[N_1/x_1, \ldots, N_n/x_n]/x_0]]\!] = \maltese$ for any $N_1 \in \mathbf{P}_1^{\perp}, \ldots, N_n \in \mathbf{P}_n^{\perp}, P \in \mathbf{N}^{\perp}$.

Clearly, $N \models \mathbf{N}$ iff $N \in \mathbf{N}$, and $P \models y : \mathbf{P}$ iff $P[x_0/y] \in \mathbf{P}$. Furthermore, by the associativity of normalization, we get the following *closure principle*:

$$P \models \mathbf{\Gamma}, x : \mathbf{P} \Longleftrightarrow \forall N \in \mathbf{P}^{\perp}, [\![P[N/x]]\!] \models \mathbf{\Gamma},$$

$$N \models \mathbf{\Gamma}, \mathbf{N} \Longleftrightarrow \forall P \in \mathbf{N}^{\perp}, [\![P[N/x_0]]\!] \models \mathbf{\Gamma}.$$

Logical Connectives and Behaviours. We next describe how behaviours are built by means of logical connectives in ludics.

An *n-ary logical connective* α is a pair $(\vec{x}_\alpha, \{a_1(\vec{x}_1), \ldots, a_m(\vec{x}_m)\})$ where $\vec{x}_\alpha = x_1, \ldots, x_n$ is a fixed sequence of variables called the *directory* of α (see [8]) and $\{a_1(\vec{x}_1), \ldots, a_m(\vec{x}_m)\}$ is a finite set of negative actions, called the *body* of the connective, such that the names $a_1, \ldots, a_m$ are distinct, the variables $\vec{x}_1, \ldots, \vec{x}_m$ are subsequences of $\vec{x}_\alpha$.

To enlighten the notation, we often identify a logical connective with its body and so in many occasions we abuse the notation, writing expression like $a(\vec{x}) \in \alpha$. Given an n-ary logical connective α and behaviours $\mathbf{N}_1, \ldots, \mathbf{N}_n$, $\mathbf{P}_1, \ldots, \mathbf{P}_n$ we define:

$$\begin{aligned} \overline{a}\langle \mathbf{N}_{i_1}, \ldots, \mathbf{N}_{i_m} \rangle &:= \{x_0 | \overline{a}\langle N_1, \ldots, N_m \rangle : N_1 \in \mathbf{N}_{i_1}, \ldots, N_m \in \mathbf{N}_{i_m}\}, \\ \overline{\alpha}\langle \mathbf{N}_1, \ldots, \mathbf{N}_n \rangle &:= \left(\textstyle\bigcup_{a(\vec{x}) \in \alpha} \overline{a}\langle \mathbf{N}_{i_1}, \ldots, \mathbf{N}_{i_m} \rangle \right)^{\perp\perp}, \\ \alpha(\mathbf{P}_1, \ldots, \mathbf{P}_n) &:= \overline{\alpha}\langle \mathbf{P}_1^{\perp}, \ldots, \mathbf{P}_n^{\perp} \rangle^{\perp}, \end{aligned}$$

where indices $i_1, \ldots, i_m$ are determined by the vector $\vec{x} = x_{i_1}, \ldots, x_{i_m}$ given for each $a(\vec{x}) \in \alpha$.

A behaviour is *logical* if it is inductively built as follows:

$$\mathbf{P} := \overline{\alpha}\langle \mathbf{N}_1, \ldots, \mathbf{N}_n \rangle, \qquad \mathbf{N} := \alpha(\mathbf{P}_1, \ldots, \mathbf{P}_n)$$

(with α an arbitrary logical connective).
Notice that the orthogonal of a logical behaviour is again logical.

Usual **MALL** connectives can be defined as follows ($*$ is a 0-ary name):

$$\begin{array}{llll} \parr := \{\wp(x_1, x_2)\}, & \otimes := \overline{\parr}, \uparrow := \{\uparrow(x_1)\}, & \perp := \{*\}, & \bullet := \overline{\wp}, \downarrow := \overline{\uparrow}, \\ \& := \{\pi_1(x_1), \pi_2(x_2)\}, & \oplus := \overline{\&}, \ \downarrow := \overline{\uparrow}, & \top := \emptyset, & \iota_i := \overline{\pi_i}. \end{array}$$

With these logical connectives we can build (semantic versions of) usual linear logic types (we use infix notations such as $\mathbf{N} \otimes \mathbf{M}$ rather than the prefix ones $\otimes\langle \mathbf{N}, \mathbf{M} \rangle$):

$\mathbf{N} \otimes \mathbf{M} = \bullet\langle \mathbf{N}, \mathbf{M} \rangle^{\perp\perp}$	$\mathbf{N} \oplus \mathbf{M} = (\iota_1\langle \mathbf{N} \rangle \cup \iota_2\langle \mathbf{M} \rangle)^{\perp\perp}$	$\downarrow \mathbf{N} = \downarrow\langle \mathbf{N} \rangle^{\perp\perp}$	$\mathbf{0} = \emptyset^{\perp\perp}$
$\mathbf{P} \parr \mathbf{Q} = \bullet\langle \mathbf{P}^{\perp}, \mathbf{Q}^{\perp} \rangle^{\perp}$	$\mathbf{P} \& \mathbf{Q} = \iota_1\langle \mathbf{P}^{\perp} \rangle^{\perp} \cap \iota_2\langle \mathbf{Q}^{\perp} \rangle^{\perp}$	$\uparrow \mathbf{P} = \downarrow\langle \mathbf{P}^{\perp} \rangle^{\perp}$	$\top = \emptyset^{\perp}$

Material and Winning Designs. Given a behaviour $\mathbf{X}$ and $D \in \mathbf{X}$, there is a "minimal portion" of D which is needed to interact with designs of $\mathbf{X}^{\perp}$. It is called *material part* of D in $\mathbf{X}$. Formally, the *intersection* $\cap$ on designs is defined by corecursion as follows:

- $P \cap \Omega = \Omega \cap P = \Omega$; $\quad \maltese \cap \maltese = \maltese$;
- $x|\overline{a}\langle \vec{N_i}\rangle \cap x|\overline{a}\langle \vec{M_i}\rangle = x|\overline{a}\langle N_i \vec{\cap} M_i\rangle$ if $N_i \cap M_i$ are defined for every $0 \leq i \leq n$;
- $\sum a(\vec{x}).P_a \cap \sum a(\vec{x}).P'_a = \sum a(\vec{x}).(P_a \cap P'_a)$ if $P_a \cap P'_a$ is defined for every $a \in A$;
- $D \cap E$ is not defined otherwise.

The material part of D in $\mathbf{X}$ is formally defined as $|D|_{\mathbf{X}} := \bigcap\{E \subseteq D : E \in \mathbf{X}\}$ and it is always a design of $\mathbf{X}$ [8, 12]. A design $D \in \mathbf{X}$ is said *material* if $D = |D|_{\mathbf{X}}$, *winning* if material and daimon-free. $|\mathbf{X}|$ (resp. $\mathbf{X}_w$) denotes the set of material (resp. winning) designs of $\mathbf{X}$.

Internal Completeness. In [8], Girard proposes a purely monistic, local notion of completeness, called *internal completeness.* It means that we can give a precise and direct description to the elements in logical behaviours without using the orthogonality and without referring to any proof system. Logical connectives easily enjoy internal completeness [12]:

- $\overline{\alpha}\langle \mathbf{N}_1, \ldots, \mathbf{N}_n\rangle = \bigcup_{a(\vec{x})\in\alpha} \overline{a}\langle \mathbf{N}_{i_1}, \ldots, \mathbf{N}_{i_m}\rangle \cup \{\maltese\}$.
- $\alpha(\mathbf{P}_1, \ldots, \mathbf{P}_n) = \{\sum a(\vec{x}).P_a : P_a \models x_{i_1} : \mathbf{P}_{i_1}, \ldots, x_{i_m} : \mathbf{P}_{i_m}$ for every $a(\vec{x}) \in \alpha\}$.

In the last equation, P_b can be arbitrary when $b(\vec{x}) \notin \alpha$. For example:

$$\mathbf{P} \,\&\, \mathbf{Q} = \{\pi_1(x_1).P + \pi_2(x_2).Q + \cdots : P \in \mathbf{P} \text{ and } Q \in \mathbf{Q}\},$$

where the non-material components of the sum are suppressed by "$\cdots$." If we consider the material designs of $\mathbf{P} \,\&\, \mathbf{Q}$, we have that $|\mathbf{P} \,\&\, \mathbf{Q}|$ is isomorphic to the *cartesian product* of $|\mathbf{P}|$ and $|\mathbf{Q}|$: a phenomenon called *mystery of incarnation* in [8].

4 An Analysis of Focalization in Ludics

Focalized Logical Behaviours. In the rest of the paper, we shall be interested in how to transform a positive logical behaviour $\mathbf{P} = \overline{\alpha}\langle\uparrow(\overline{\beta}\langle \mathbf{N}_1, \ldots \mathbf{N}_m\rangle), \mathbf{M}_2, \ldots, \mathbf{M}_n\rangle$ into a behaviour $\mathbf{Q} = \overline{\alpha\beta}\langle \mathbf{N}_1, \ldots, \mathbf{N}_m, \mathbf{M}_2, \ldots, \mathbf{M}_n\rangle$, where $\mathbf{M}_i, \mathbf{N}_j$ are negative logical behaviours and $\alpha = (\vec{x}_\alpha, \{a_1(\vec{x}_1), a_2(\vec{x}_2), \ldots\})$ with $\vec{x}_\alpha = x_1, \ldots, x_n$, $\beta = (\vec{y}_\beta, \{b_1(\vec{y}_1), b_2(\vec{y}_2), \ldots\}$ with $\vec{y}_\beta = y_1, \ldots, y_m$ such that $\vec{x}_\alpha$ and $\vec{y}_\beta$ are *disjoint.* $\mathbf{Q}$ is called the *focalized behaviour* associated to $\mathbf{P}$ (relative to α, β) while $\alpha\beta$ is the *synthetic connective* associated to α, β.

The choice of having $\uparrow(\overline{\beta}\langle \mathbf{N}_1, \ldots \mathbf{N}_m\rangle)$ as $\mathbf{M}_1$ and not, for example as $\mathbf{M}_j$, is of course completely arbitrary and aims at making the presentation simpler. On the other hand, while $\mathbf{M}_2, \dots, \mathbf{M}_n$ are arbitrary, $\uparrow(\overline{\beta}\langle \mathbf{N}_1, \dots \mathbf{N}_m\rangle)$ has always *this special form*, with the negative connective $\uparrow$ as prefix: focalization roughly asserts that such dummy actions occurring in designs of $\mathbf{P}$ can always be removed by considering synthetic connectives.

In the remaining of this section, and unless otherwise stated, $\mathbf{P}$ and $\mathbf{Q}$ will respectively denote $\overline{\alpha}\langle \uparrow(\overline{\beta}\langle \mathbf{N}_1, \dots \mathbf{N}_m\rangle), \mathbf{M}_2, \dots, \mathbf{M}_n\rangle$ and $\overline{\alpha\beta}\langle \mathbf{N}_1, \dots, \mathbf{N}_m, \mathbf{M}_2, \dots, \mathbf{M}_n\rangle$.

Synthetic Connectives. In order to define the focalized behaviour $\mathbf{Q}$ we shall define properly the synthetic connective $\alpha\beta$, by specifying its directory and its body:

- The directory of $\alpha\beta$ is $\vec{z}_{\alpha\beta} := y_1, \dots, y_m, x_2, \dots, x_n$. Hence, $\alpha\beta$ has arity $n+m-1$.
- The body of $\alpha\beta$ consists of the set of negative actions $ab(\vec{z})$ defined as follows. First notice that our definition of logical connectives ensures that if some action $a(x_{a_1}, \dots, x_{a_{k_a}})$ in α is such that $x_1 \in x_{a_1}, \dots, x_{a_{k_a}}$, then $x_1 = x_{a_1}$. Thus, for any $a(x_{a_1}, \dots, x_{a_{k_a}})$ in the body of α and $b(y_{b_1}, \dots, y_{b_{l_b}})$ in the body of β, we define a new action ab as:

$$\begin{array}{rl} ab(x_{a_1}, \dots, x_{a_{k_a}}) & \text{if } x_1 \notin x_{a_1}, \dots, x_{a_{k_a}}, \\ ab(y_{b_1}, \dots, y_{b_{l_b}}, x_{a_2}, \dots, x_{a_{k_a}}) & \text{if } x_1 = x_{a_1}. \end{array}$$

To sum up, we can associate to $\overline{\alpha}\langle \uparrow(\overline{\beta}\langle \mathbf{N}_1, \dots \mathbf{N}_m\rangle), \mathbf{M}_2, \dots, \mathbf{M}_n\rangle$ its focalized behaviour (relative to α, β) $\overline{\alpha\beta}\langle \mathbf{N}_1, \dots \mathbf{N}_m, \mathbf{M}_2, \dots, \mathbf{M}_n\rangle$. The following examples illustrate this:

(a) Let $\mathbf{P}$ be $\otimes\langle \uparrow(\downarrow\langle \mathbf{N}\rangle), \mathbf{M}\rangle$ (written as $\uparrow\downarrow\mathbf{N} \otimes \mathbf{M}$ in infix notation). Since $⅋$ and $\uparrow$ are respectively $(x_1x_2, \{\wp(x_1, x_2)\})$ and $(y, \{\uparrow(y)\})$ with x_1, x_2, y distinct, we have $⅋\uparrow = (\{yx_2, \wp\uparrow(y, x_2)\})$ and $\mathbf{Q} = \overline{⅋\uparrow}\langle \mathbf{N}, \mathbf{M}\rangle = \otimes\downarrow\langle \mathbf{N}, \mathbf{M}\rangle$. Note that $\otimes\downarrow$ and $\otimes$ are isomorphic.

(b) Let $\mathbf{P}$ be $\oplus\langle \uparrow(\otimes\langle \mathbf{N}_1, \mathbf{N}_2\rangle), \mathbf{M}\rangle$ (written $\uparrow(\mathbf{N}_1 \otimes \mathbf{N}_2) \oplus \mathbf{M}$ in infix notation). Since $\&$ and $⅋$ are respectively $(x_1x_2, \{\pi_1(x_1), \pi_2(x_2)\})$ and $(y_1y_2, \{\wp(y_1, y_2)\})$ we have that $\&⅋ = (y_1y_2x_2, \{\pi_1\wp(y_1, y_2), \pi_2\wp(x_2)\})$ and finally $\mathbf{Q} = \overline{\&⅋}\langle \mathbf{N}_1, \mathbf{N}_2, \mathbf{M}\rangle = \oplus\otimes\langle \mathbf{N}_1, \mathbf{N}_2, \mathbf{M}\rangle$. Notice that in this case $\pi_2\wp(x_2)$ is just $\pi_2(x_2)$, with an irrelevant change of name.

Now we show how to obtain $\mathbf{Q}$ from $\mathbf{P}$ *interactively*, by means of *interactive functions*.

Interactive Functions. Given two positive (resp. negative) logical behaviours $\mathbf{F}$, $\mathbf{G}$, an *interactive function* (i–function for short) $f : \mathbf{F} \longrightarrow \mathbf{G}$ is any design $f \models \mathbf{F}^{\perp}, x_0 : \mathbf{G}$ (resp. $f \models \mathbf{G}, x_0 : \mathbf{F}^{\perp}$). We write $f(P)$ for $[\![P[f/x_0]]\!]$ if $P \in \mathbf{F}$ (resp. $f(M)$ for $[\![f[M/x_0]]\!]$ if $M \in \mathbf{F}$) and f a i–function. We also write $f(\mathbf{F})$ for $\{f(D) : D \in \mathbf{F}\}$. Observe that since our setting is fully linear, i–functions have to be intended as "linear" functions. Two examples follow:

(a) A very important i–function is the *fax* [8] (or η-expanded identity) recursively defined as $\mathsf{i}(x_0) := \sum \mathsf{i}(x_0)_a$ with $\mathsf{i}(x_0)_a := a(y_1, \dots, y_k).x_0|\overline{a}\langle \mathsf{i}(y_1), \dots \mathsf{i}(y_k)\rangle$ in [12].

$\mathsf{i}(x_0)$ plays the role of the identity function for designs: $\mathsf{i}(x_0)(D) = D$ for any D.

(b) We define $\mathsf{u}_{\alpha\beta} : \mathbf{Q} \longrightarrow \mathbf{P}$ as $\mathsf{u}_{\alpha\beta} := \sum_{\alpha\beta} \mathsf{u}_{ab} + \sum_{c \notin \alpha\beta} \mathsf{i}(x_0)_c$ with u_{ab}, for any $ab \in \alpha\beta$, defined as (abbreviating $y_{b_1}, \ldots, y_{b_{l_b}}$ by $\mathbf{y}$ and $\mathsf{i}(y_{b_1}), \ldots, \mathsf{i}(y_{b_{l_b}})$ by $\mathsf{i}(\mathbf{y})$):

$$\mathsf{u}_{ab} := ab(x_{a_1}, \ldots, x_{a_{k_a}}).x_0|\overline{a}\langle \mathsf{i}(x_{a_1}), \ldots, \mathsf{i}(x_{a_{k_a}})\rangle \quad \text{if } x_1 \neq x_{a_1}$$
$$\mathsf{u}_{ab} := ab(\mathbf{y}, x_{a_2} \ldots, x_{a_{k_a}}).x_0|\overline{a}\langle \uparrow(y).y|\overline{b}\langle \mathsf{i}(\mathbf{y})\rangle, \mathsf{i}(x_{a_2}), \ldots, \mathsf{i}(x_{a_{k_a}})\rangle \quad \text{if } x_1 = x_{a_1}.$$

$\mathsf{u}_{\alpha\beta}$, which sends designs in $\mathbf{Q}$ to designs in $\mathbf{P}$, will be important in analyzing the interactive focalization process of the *focalizing-design* f. The role of $\mathsf{u}_{\alpha\beta}$ is to break a synthetic connective $\alpha\beta$ into its more atomic connectives α and β.

Section-Retraction Pairs. Given two logical behaviours of the same polarity $\mathbf{F}, \mathbf{G}$, a *section-retraction pair* from $\mathbf{G}$ to $\mathbf{F}$ is a pair of i-functions (s, r) with $s : \mathbf{G} \longrightarrow \mathbf{F}$, the section, and $r : \mathbf{F} \longrightarrow \mathbf{G}$, the retraction, such that $r \circ s = [\![s[r/x_0]]\!] = \mathsf{i}(x_0)$. A section-retraction pair is *strict* if it sends a daimon-free design to a daimon-free one. Section-retraction pairs can be considered in a context:

Theorem 1. *Any (strict) section-retraction pairs between* $\mathbf{P}_i$ *and* $\mathbf{Q}_i$ *(*$i = 1, \ldots, n$*) can be extended to a (strict) section-retraction pair between* $\alpha(\mathbf{P}_1, \ldots, \mathbf{P}_n)$ *and* $\alpha(\mathbf{Q}_1, \ldots, \mathbf{Q}_n)$ *for any logical connective* α*. The same holds for the positive case.*

Then, focalization can be expressed as the existence of a section-retraction pair from $\mathbf{Q}$ to $\mathbf{P}$ with $\mathsf{u}_{\alpha\beta}$ as section.

The Focalizing-Design $\mathsf{f}_{\alpha\beta}$. We now introduce the i–function $\mathsf{f}_{\alpha\beta} : \mathbf{P} \longrightarrow \mathbf{Q}$, which will be the retraction associated with $\mathsf{u}_{\alpha\beta}$ and shall interactively build the *focalized designs.* $\mathsf{f}_{\alpha\beta}$ is defined as $\mathsf{f}_{\alpha\beta} := \sum_{\alpha\beta} \mathsf{f}_{ab} + \sum_{c \notin \alpha\beta} \mathsf{i}(x_0)_c$ with, for any $ab \in \alpha\beta$, f_{ab} being defined as:

$$\mathsf{f}_{ab} := a(x_{a_1}, \ldots, x_{a_{k_a}}).x_0|\overline{ab}\langle \mathsf{i}(x_{a_1}), \ldots, \mathsf{i}(x_{a_{k_a}})\rangle \quad \text{if } x_1 \neq x_{a_1},$$
$$\mathsf{f}_{ab} := a(x_1, x_{a_2}, \ldots, x_{a_{k_a}}).x_1|\downarrow\langle \textstyle\sum_\beta b(\mathbf{y}).x_0|\overline{ab}\langle \mathsf{i}(\mathbf{y}), \mathsf{i}(x_{a_2}), \ldots, \mathsf{i}(x_{a_{k_a}})\rangle\rangle \quad \text{if } x_1 = x_{a_1}.$$

Theorem 2

1. $\mathsf{f}_{\alpha\beta}(\mathbf{P}) = \mathbf{Q}$ *and winning conditions are preserved:* $\mathsf{f}_{\alpha\beta}(\mathbf{P}_w) \subseteq \mathbf{Q}_w$ *(actually,* $\mathsf{f}_{\alpha\beta}(\mathbf{P}_w) = \mathbf{Q}_w$*).*
2. $\mathsf{u}_{\alpha\beta}(|\mathbf{Q}|) = |\mathbf{P}|$ *and winning conditions are preserved:* $\mathsf{u}_{\alpha\beta}(\mathbf{Q}_w) \subseteq \mathbf{P}_w$*.*

Composing $\mathsf{f}_{\alpha\beta}$ and $\mathsf{u}_{\alpha\beta}$. To establish that $(\mathsf{u}_{\alpha\beta}, \mathsf{f}_{\alpha\beta})$ is a section-retraction pair from $\mathbf{Q}$ to $\mathbf{P}$, we shall study the *composition* of the two i-functions $\mathsf{f}_{\alpha\beta} \circ \mathsf{u}_{\alpha\beta}$. We have:

Proposition 1. $\mathsf{f}_{\alpha\beta} \circ \mathsf{u}_{\alpha\beta} = \mathsf{i}(x_0)$.

Proof. By definition of $\mathsf{f}_{\alpha\beta}$ and $\mathsf{u}_{\alpha\beta}$, it is immediate that

$$\mathsf{f}_{\alpha\beta} \circ \mathsf{u}_{\alpha\beta} = [\![\mathsf{u}_{\alpha\beta}[\mathsf{f}_{\alpha\beta}/x_0]]\!] = [\![\sum_{\alpha\beta} \mathsf{u}_{ab}[\sum_{\alpha\beta} \mathsf{f}_{ab}/x_0]]\!] + \sum_{c \notin \alpha\beta} \mathsf{i}(x_0)_c.$$

Moreover, since $[\![\sum_{\alpha\beta} \mathsf{u}_{ab}[\sum_{\alpha\beta} \mathsf{f}_{ab}/x_0]]\!] = \sum_{\alpha\beta}[\![\mathsf{u}_{ab}[\mathsf{f}_{ab}/x_0]]\!]$, the left member of the sum can be further decomposed and we have two cases: if $ab(\vec{z})$ is $ab(x_{a_1}, \dots, x_{a_{k_a}})$, we have that:

$$\begin{aligned}
[\![\mathsf{u}_{ab}[\mathsf{f}_{ab}/x_0]]\!] &= ab(x_{a_1}, \dots, x_{a_{k_a}}).[\![\mathsf{f}_{ab}|\overline{a}\langle \mathsf{i}(x_{a_1}), \dots, \mathsf{i}(x_{a_{k_a}})\rangle]\!] \\
&= ab(x_{a_1}, \dots, x_{a_{k_a}}).x_0|\overline{ab}\langle [\![\mathsf{i}(\mathsf{i}(x_{a_1}))]\!], \dots, [\![\mathsf{i}(\mathsf{i}(x_{a_{k_a}}))]\!]\rangle \\
&= ab(x_{a_1}, \dots, x_{a_{k_a}}).x_0|\overline{ab}\langle \mathsf{i}(x_{a_1}), \dots, \mathsf{i}(x_{a_{k_a}})\rangle \;=\; \mathsf{i}(x_0)_{ab}.
\end{aligned}$$

Otherwise, if $ab(\vec{z}) = ab(y_{b_1}, \dots, y_{b_{l_b}}, x_{a_2} \dots, x_{a_{k_a}})$, writing $\mathbf{x}$ for $x_{a_2}, \dots, x_{a_{k_a}}$, we have that

$$\begin{aligned}
[\![\mathsf{u}_{ab}[\mathsf{f}_{ab}/x_0]]\!] &= ab(\mathbf{y}, \mathbf{x}).[\![f_{ab}|\overline{a}\langle \uparrow(y).y|\overline{b}\langle \mathsf{i}(\mathbf{y})\rangle, \mathsf{i}(\mathbf{x})\rangle]\!] \\
&= ab(\mathbf{y}, \mathbf{x}).[\![(\uparrow(y).y|\overline{b}\langle \mathsf{i}(\mathbf{y})\rangle)|\downarrow\langle \sum_\beta b(\mathbf{y}).x_0|\overline{ab}\langle \mathsf{i}(\mathbf{y}), \mathsf{i}(\mathsf{i}(\mathbf{x}))\rangle\rangle]\!] \\
&= ab(\mathbf{y}, \mathbf{x}).[\![(\sum_\beta b(\mathbf{y}).x_0|\overline{ab}\langle \mathsf{i}(\mathbf{y}), \mathsf{i}(\mathsf{i}(\mathbf{x}))\rangle)|\overline{b}\langle \mathsf{i}(\mathbf{y})\rangle]\!] \\
&= ab(\mathbf{y}, \mathbf{x}).x_0|\overline{ab}\langle [\![\mathsf{i}(\mathsf{i}(\mathbf{y}))]\!], [\![\mathsf{i}(\mathsf{i}(\mathbf{x}))]\!]\rangle \\
&= ab(\mathbf{y}, \mathbf{x}).x_0|\overline{ab}\langle \mathsf{i}(\mathbf{y}), \mathsf{i}(\mathbf{x})\rangle \\
&= \mathsf{i}(x_0)_{ab}.
\end{aligned}$$

Finally, we have obtained that $[\![\mathsf{f}_{\alpha\beta}(\mathsf{u}_{\alpha\beta})]\!] = \mathsf{i}(x_0)$.

Focalization Theorem. We can now conclude with the focalization theorem:

Theorem 3 (Focalization Theorem). *For any logical connectives α and β, there is a strict section-retraction pair from $\overline{\alpha\beta}\langle \mathbf{Y}, \mathbf{X}\rangle$ to $\overline{\alpha}\langle \uparrow(\overline{\beta}\langle \mathbf{Y}\rangle), \mathbf{X}\rangle$ which is the pair $(\mathsf{u}_{\alpha\beta}, \mathsf{f}_{\alpha\beta})$.*

An important thing to notice is that Theorem 1 applies to $(\mathsf{u}_{\alpha\beta}, \mathsf{f}_{\alpha\beta})$ and that the section-retraction pair is strict. This will allow to carry the building of synthetic connectives inside contexts and to ensure that we will obtain proofs via full completeness. $\mathsf{f}_{\alpha\beta}$ is thus a retraction from $\mathbf{P}$ to $\mathbf{Q}$ which will map proofs to proofs with synthetic connectives. Moreover, $\mathsf{u}_{\alpha\beta} \circ \mathsf{f}_{\alpha\beta} : \mathbf{P} \longrightarrow \mathbf{P}$ is an interactive function from $\mathbf{P}$ to $\mathbf{P}$ which preserves winning conditions: given a proof (winning design), it shall build a focussing version of that proof.

5 Conclusion and Future Works

We have considered in this short abstract how focalization can be considered from the point of view of ludics itself. In order to do so, we considered interactive functions which have the ability to make a cluster of two positive logical connectives which are separated by a single trivial $\uparrow$ logical connective (that is to merge them in a single synthetic connective), while preserving winning conditions.

Our present work naturally leads to directions that we shall develop in future works:

- A natural direction is to get a more semantical proof of the focalization theorem for **MALL** by combining the results in the present paper with the full completeness results of ludics [8].
- We would like to extend our results to the case of the exponentials [2, 3], not only because our current analysis is restricted to the linear case, but also because it might clarify several elements of the proof–theory of the exponentials (and their bipolar behaviour).
- The initial motivation of our work was to find an analogous to the tape compression theorem for Turing machines. We also plan to develop this line of work in the future.

References

[1] Andreoli, J.-M.: Logic Programming with Focusing Proofs in Linear Logic. Journal of Logic and Computation 2(3), 297–347 (1992)
[2] Basaldella, M., Faggian, C.: Ludics with Repetitions (Exponentials, Interactive Types and Completeness). In: LICS (2009)
[3] Basaldella, M., Terui, K.: On the meaning of logical completeness. In: Curien, P.-L. (ed.) TLCA 2009. LNCS, vol. 5608, pp. 50–64. Springer, Heidelberg (2009)
[4] Curien, P.-L.: Introduction to LL and ludics, I & II. In: Advances in Mathematics, China, vol. 34(5), pp. 513–544 (2005); vol. 35(1), pp. 1–44 (2006)
[5] Faggian, C., Piccolo, M.: Ludics is a model for the finitary linear pi-calculus. In: Della Rocca, S.R. (ed.) TLCA 2007. LNCS, vol. 4583, pp. 148–162. Springer, Heidelberg (2007)
[6] Girard, J.-Y.: On the meaning of logical rules I: syntax vs. semantics. In: Berger, U., Schwichtenberg, H. (eds.) Computational Logic, pp. 215–272. Springer, Heidelberg (1999)
[7] Girard, J.-Y.: On the meaning of logical rules II: multiplicatives and additives. In: Berger, Schwichtenberg (eds.) Foundation of Secure Computation, pp. 183–212 (2000)
[8] Girard, J.-Y.: Locus solum: From the rules of logic to the logic of rules. Mathematical Structures in Computer Science 11, 301–506 (2001)
[9] Laurent, O.: Étude de la polarization en logique. PhD thesis, Univ. Aix-Marseille II (2002)
[10] Laurent, O.: A proof of the focalization property of Linear Logic (2004) (unpublished note)
[11] Miller, D., Saurin, A.: From proofs to focused proofs: A modular proof of focalization in linear logic. In: Duparc, J., Henzinger, T.A. (eds.) CSL 2007. LNCS, vol. 4646, pp. 405–419. Springer, Heidelberg (2007)
[12] Terui, K.: Computational ludics. To appear in Theoretical Computer Science (2008)

On Some Logic Games in Their Philosophical Context

Tero Tulenheimo*

STL-CNRS / University of Lille 3

Abstract. In the present paper Hintikka's game-theoretical semantics and the dialogical logic of Lorenzen and Lorenz are discussed and compared from the viewpoint of their underlying philosophical meaning theories. The question of whether the proposed meaning theories can be claimed to suffer from circularity is taken up. The relations of the two frameworks to verificationist and anti-realist ideas are considered. Finally, van Heijenoort's concept of 'logic as calculus' generalized by Hintikka to the idea of 'language as calculus' will be reformulated as a view we label 'anti-universalism.' We discuss briefly the fourfold division of foundational views obtained by relating a distinction between 'universalism' and 'anti-universalism' to the distinction between 'realism' and 'anti-realism.'

1 Introduction

The topic 'logic and games' is *en vogue* these days, on many fronts. Due to their interest in interactive computation, various researchers have been led to utilize games in connection with logic for different purposes: game semantics of programming languages (Abramsky & Jagadeesan & Malacaria, Hyland & Ong), ludics (Girard) and computability logic (Japaridze), to name some of the best-known examples. A host of others (notably van Benthem and his collaborators) seek to see logic from a generalized viewpoint as a study of rational agency and intelligent interaction. The latter agenda in principle subsumes the former, but in practice the research questions that emerge from van Benthem's approach use classical logic and model-theoretical tools whereas those interested in interactive computation typically have a proof-theoretical inclination and take inspiration from intuitionistic logic. Also, representatives of the former agenda think of semantics as associating certain sorts of actions or processes with logical operators, in contrast to the latter framework where model-theoretical truth conditions are usually used for describing whatever sorts of actions or processes one happens to be interested in[1]. There are, then, essential differences in the

* The research for the present paper was partially funded by the Academy of Finland (project no. 207188, 'Modalities, Games and Independence in Logic'), and partially supported by a personal grant from Ella and Georg Ehrnrooth foundation. I wish to thank Shahid Rahman and Helge Rückert for helpful discussions, and the anonymous referee for useful comments.

[1] For representatives of the two agendas, see [1, 4, 9, 25, 26].

Lecomte and S. Tronçon (Eds.): PRELUDE 2010, LNAI 6505, pp. 88–113, 2011.

approaches of the researchers in the area of 'logic and games' — which is not surprising, as related 'foundational' or 'motivational' differences have existed at least ever since the contrast between classical and intuitionistic logic became a theme for philosophical reflection[2].

The present paper does *not* discuss the two above-mentioned contemporary trends in the study of 'logic and games.' Rather, the two main historical precursors of this field are discussed with reference to the philosophical ideas that underlie them: Lorenzen's dialogical logic and Hintikka's game-theoretical semantics (or GTS)[3]. Both approaches associate games with logical formulas in such a way that the semantic attributes of interest become definable via the existence of a winning strategy for a suitable player, wherefore they both may be said to offer, each in its own way, a *game-theoretical semantics*. In referring to Hintikka's approach, 'game-theoretical semantics' has come to be used as a proper name. Since Lorenzen's approach can be termed 'dialogical logic' or be said to provide a 'dialogue semantics,' in the interest of clarity we avoid applying the common noun 'game-theoretical semantics' when speaking of it. This decision is reasonable not only to avoid confusion but also because the two frameworks represent, as will be seen, very different views on what counts as semantics in the first place. The choice is not intended to suggest any priorities. Lorenzen enjoys the historical priority of having been, with his 1958 talk 'Logik und Agon' (published in 1960), the first in the 20th century who began to draw systematic and philosophically motivated connections between logic and games. Hintikka's ideas were first expounded in his John Locke lectures at Oxford in 1964; the first publication, 'Language-Games for Quantifiers,' appeared in 1968. Incidentally, Lorenzen too delivered John Locke lectures at Oxford, during the academic year 1968–1969[4].

We begin by sketching the basic ideas of dialogical logic on the one hand and GTS on the other (Sections 2 and 3). These approaches are then compared for their underlying philosophical meaning theories. Their differences are stressed. Potentially problematic ideas in both approaches will be detected (Section 4). We consider the relations of the two frameworks to verificationist and anti-realist ideas (Section 5). Finally, van Heijenoort's idea of 'logic as calculus' generalized by Hintikka to the idea of 'language as calculus' will be reformulated as a view to be labeled 'anti-universalism.' We discern Hintikka's and Lorenzen's positions in the fourfold division of foundational views induced by the distinctions realism/anti-realism and universalism/anti-universalism (Section 6).

Throughout the paper we confine our discussion to *logical expressions*, more specifically to the logical operators of first-order logic, at the expense of lexical items and, generally, expressions and constructions appearing in natural

---

[2] Of course in many concrete cases one's 'foundational ideas' are imposed by the research tradition one happens to find oneself in, rather than being results of any sophisticated weighing of the options.

[3] For a discussion, see [37, 42, 48] (dialogues), [5, 22, 39] (GTS), [46, 50] (comparing the two).

[4] For the respective lectures in published form, see [12] and [33].

languages[5]. In connection with logical expressions, both approaches work on two fronts: on the one hand their goal is to explicate how these expressions receive their *meaning*, and on the other hand they aim to indicate how the relevant *semantic attributes* emerge. For GTS, the attribute of interest is *(material) truth*. For the dialogian it can be truth as well as *validity* — which one, depends on the sort of dialogue at hand. Often people discussing dialogues are exclusively interested in the so-called formal dialogues; *they* are preoccupied with the semantic attribute of validity.

2 Dialogical Logic

The basic idea of the dialogical approach is that meanings of expressions are always connected with actions. The specific meaning that an expression has is determined by the sorts of norms or rules to which the use of this expression is subjected. One may discern rules of more than one type, but all the rules are, one could say, *argumentative* or *discursive*. The basic unit on which the rules operate is an *utterance* or *assertion* (Behauptung, Aussage), not an expression or a sentence in isolation from the act of uttering it. Dialogue rules taken jointly serve to pin down a dialogue context in which utterances of the relevant language may be evaluated. These rules can be seen as determining a 'language game' or a linguistic practice — or perhaps more specifically an argumentative practice. There is in principle no claim that one could not choose one's rules in different ways, obtaining different linguistic practices even for the same logical expressions. In practice, though, Lorenzen himself took one set of rules to be fundamental in connection with logic (namely the one leading to 'intuitionistic dialogues,' to be described below).

Among dialogue rules it is customary to distinguish between *particle rules* (Partikelregeln) and *structural rules* (Rahmenregeln). The former, a.k.a. attack-defense rules, incorporate the idea that utterances induce commitments. The rules are normative: they tell how such commitments may be tested, and specify the utterer's obligations triggered by a given test. To express it otherwise, the particle rules indicate how an utterance may be attacked and how one may defend one's utterance against a given attack. If X and Y are distinct players, the rules are often given schematically along the following lines:

(R.∧):

utterance	$X : (A \wedge B)$
attack	$Y : ?_L$ or $Y : ?_R$
defense	$X : A$ *resp.* $X : B$

(R.∨):

utterance	$X : (A \vee B)$
attack	$Y : ?_\vee$
defense	$X : A$ or $X : B$

(R.→):

utterance	$X : (A \rightarrow B)$
attack	$Y : A$
defense	$X : B$

(R.¬):

utterance	$X : \neg A$
attack	$Y : A$
defense	$X : —$

[5] For GTS and natural languages, see [19, 20, 21]. For Lorenzen's views, see e.g. [27, 32, 34].

(R.∀):

utterance	$X : \forall x A$
attack	$Y : ?_{\mathrm{t}}$
defense	$X : A[x/\mathrm{t}]$

(R.∃):

utterance	$X : \exists x A$
attack	$Y : ?_{\exists}$
defense	$X : A[x/\mathrm{t}]$

In the quantifier rules, t is a term taken from some stock of terms assumed to be available[6]. The rule $(R.\neg)$ states that no response to an attack on $\neg A$ is possible. (Insofar as the attack is concerned, this leaves for X only the possibility to present a counterattack, i.e., to attack against Y's utterance of A; intuitively such a counterattack should be successful if $\neg A$ is defensible as a thesis stated by X.) Particle rules are meant to lay down local instructions for how to act: at any position the players have reached in the course of a dialogue, these rules tell how a sentence that has already been uttered can be attacked or defended — if no further rule blocks it from being so processable.

Structural rules complement the particle rules by a sufficient number of additional stipulations so as to determine in detail how a dialogue can be conducted. Jointly these rules specify, *for each sentence* A of the language considered, how the dialogue $\mathcal{D}(A)$ about A is played[7]. The sentence A is termed the *thesis* of the dialogue $\mathcal{D}(A)$. Dialogues have two players or dialogue partners X and Y, one of whom adopts the 'role' of proponent and the other that of opponent; for simplicity the players are referred to via their roles — as **P** and **O**[8]. We say that Y is the 'adversary' of X and *vice versa*. Here is a formulation of the structural rules for *formal* dialogues about first-order sentences:

1. **Starting rule:** The initial move consists of **P**'s uttering the thesis, if possible. (Rule 5 prevents him from doing so if the thesis is atomic.) If **P** has made the initial move, player **O** chooses a natural number n and then player **P** chooses a natural number m. (Rule 2 will make use of these numbers.) Thereafter the players move alternately, each move after the initial move being an attack or a defense.
2. **Repetition rule:** The numbers n and m chosen initially are termed the *repetition ranks* of **O** and **P**, respectively[9]. In the course of the dialogue, **O** (**P**) may attack or defend any single (token of an) utterance at most n (respectively m) times.
3. **Winning rule:** Whoever cannot move has *lost* and his or her adversary has *won*.
4. (a) **Intuitionistic rule:** Each player may attack any complex sentence uttered by the adversary, or respond to *the last attack (against one of his or her earlier utterances) to which no defense has yet been presented.* That

[6] $A[x/\mathrm{t}]$ is the result of replacing all free occurrences of x in A by t.

[7] We define dialogues for sentences only, though they could be formulated for arbitrary formulas.

[8] These 'roles' are in effect defined by the *starting rule* (see below): the player who is supposed to move first is **P**, the other player is **O**. Let us agree that **O** is female ('she') and **P** male ('he').

[9] For attack and defense ranks, cf. [28, pp. 40, 84–85, 98–99], [33, p. 28]. The notion of repetition rank subsumes both attack and defense ranks.

is, the move that has been attacked last must be defended first. One may never revise an earlier defense, and in general one may not postpone a defense indefinitely without losing the possibility of that defense. (One might return to respond to an earlier attack only by first successfully responding to all intervening attacks.)

(b) **Classical rule:** Each player may attack any complex sentence uttered by the adversary, or respond to *any* attack against his or her earlier utterance, including those that have already been defended. That is, not only can one postpone responses to attacks indefinitely without endangering the possibility to come back to them, but also revising old defenses is possible.

5. **Formal rule:** Player **P** may not utter an atomic sentence unless it has already been uttered by **O**. We may say that **P**'s utterance of an atomic sentence must be *grounded* by **O**'s previous utterance of the same atomic sentence.

The particle rules combined with the structural rules (1), (2), (3), (4a) and (5) give rise to *intuitionistic formal dialogues.* To obtain *classical formal dialogues*, rule (4a) is replaced by rule (4b). The so-called *material dialogues* — which will not be systematically discussed in the present paper — presuppose that (at least implicitly) a model determining the truth-values of atomic sentences be given. Repetition ranks may be infinite ordinal numbers if the domain of the model is infinite. The winning rule says that whoever utters a false atomic sentence, or cannot move, has lost while his or her adversary has won. Material dialogues have no formal rule and, as a matter of fact, in connection with them it makes no difference whether the intuitionistic rule (4a) or the classical rule (4b) is adopted[10].

The players are assumed to agree on criteria for settling whether the rules are followed, whether a play has come to an end, and who in that case has won[11]. The dialogical criteria for the 'correctness' of an utterance are ultimately intersubjective; the final arbiter is mutual agreement, rather than an independently existing reality as such. Any sequence of moves made in accordance with the dialogue rules (i.e., particle rules and structural rules) is a *partial play.* A *play* is a partial play at which the player whose turn it is to move is incapable of doing so. A *strategy* for player Z is a set of instructions which for every partial play at which it is Z's turn to move indicates, if possible, a move complying with the rules. A strategy for Z is *winning* if against any sequence of moves by Z's adversary, it yields moves so that the resulting play is won by Z.

The dialogicians take the *meanings* of the logical operators to be given by the dialogue rules — in such a way that the particle rules contribute some sort of core meaning or local meaning of the logical operators, thought of as remaining intact when structural rules are varied. *Semantic attributes*, again, emerge with

[10] For a comparison between material dialogues and the semantic games of GTS, see [52].

[11] Cf. [28, p. 35] and [31, p. 21]. These requirements introduce an anti-realist element into the dialogical framework.

reference to the notion of winning strategy. Meaning is determined by fixing the level of plays, while semantic attributes are determined at the strategic level. Quite generally, the thesis A of the dialogue $\mathcal{D}(A)$ has one semantic attribute when there exists a winning strategy for **P** in $\mathcal{D}(A)$, and it has another semantic attribute when it is **O** for whom there is a winning strategy[12]. E.g., the dialogician might *define* the validity (refutability) of sentence A of intuitionistic first-order logic as the existence of a winning strategy for **P** (respectively, **O**) in the formal intuitionistic dialogue about A. Similarly, material truth (falsity) of a sentence can be defined as the existence of a winning strategy for **P** (respectively, **O**) in the material dialogue about that sentence.

Let us take a couple of examples of plays, starting with formal dialogues. The numbers in the margins indicate the order in which the moves are made. Numbers near the middle divide appear only in connection with attacks and specify to which utterance by the adversary the attack pertains. We write an attack and the reply to *that* attack, if any, always on the same row.

	O			P	
				$A \vee \neg A$	0
1	$n := 1$			$m := 2$	2
3	$?_{\vee}$	0		$\neg A$	4
5	A	4		—	
				A (revising the defense against move 3)	6

	O			P	
				$A \vee \neg A$	0
1	$n := 1$			$m := 2$	2
3	$?_{\vee}$	0		$\neg A$	4
5	A	4		—	

Fig. 1. Classical play won by **P** (left); intuitionistic play won by **O** (right)

Player **P** can force a win in the classical dialogue about $A \vee \neg A$ by choosing $m \geq n + 1$. In the corresponding intuitionistic dialogue, it is **O** who can force a win: she may simply choose $n := 1$. The intuitionistic rule blocks **P**'s possibility of revising the defense of the disjunction (since **P** would first need to reply to **O**'s attack against the negation $\neg A$ which is impossible); and there are no attacks for him to repeat. For the following example, let B be a formula in which the variable x does not appear free, e.g., $P(\mathsf{c})$.

	O			P	
				$\forall x(B \vee C) \to (B \vee \forall x C)$	0
1	$n := 1$			$m := 1$	2
3	$\forall x(B \vee C)$	0		$B \vee \forall x C$	4
5	$?_{\vee}$	4		$\forall x C$	6
7	$?_{\mathsf{t}}$	6			
9	$B \vee C[x/\mathsf{t}]$		3	$?_{\mathsf{t}}$	8
11	B		9	$?_{\vee}$	10

Fig. 2. Intuitionistic play won by **O**

[12] Formal dialogues for first-order logic are *well-founded* (each play is finitely long) games of *perfect information* (each player is fully informed of all the past moves). By the Gale-Stewart theorem, any such dialogue is *determined*: there is a winning strategy for one or other of the players.

In the above play there are no more defensive moves available for **P**, because the first such move would have to be a response to the attack of move 7 to which he cannot respond, and he has furthermore already attacked once all moves available for attack. There is actually a winning strategy for **O** in the intuitionistic dialogue about $\forall x(B \vee C) \rightarrow (B \vee \forall xC)$. It suffices that she chooses $n := 1$, takes care not to introduce B before **P** has uttered $\forall xC$, attacks **P**'s utterance of $\forall xC$ as soon as possible by some term k and then takes care of never introducing $C[x/\mathsf{k}]$, which is possible since B can be introduced instead. (It is not difficult to see that **P** cannot improve his payoff by choosing his repetition rank to be greater than 1.)

Let us then illustrate material dialogues. Let $\mathcal{M}$ be a model of vocabulary $\{R, \mathsf{t}_1, \mathsf{t}_2\}$ making both atomic sentences $R\mathsf{t}_2\mathsf{t}_1$ and $R\mathsf{t}_2\mathsf{t}_2$ true[13]. Further, let $\mathcal{N}$ be a model of vocabulary $\{P, Q, \mathsf{t}_1, \mathsf{t}_2, \mathsf{t}_3, \ldots\}$ in which all of the following atomic sentences are true: $P\mathsf{t}_7$ and $Q\mathsf{t}_1$, $Q\mathsf{t}_2$, $Q\mathsf{t}_3$, etc.

O		**P**		
			$\neg\exists x\forall yRxy$	0
1	$n := 1$		$m := 2$	2
3	$\exists x\forall yRxy$ 0		—	
5	$\forall yR\mathsf{t}_2y$	3	$?_\exists$	4
7	Rt_2t_1	5	$?_{t_1}$	6
9	Rt_2t_2	5	$?_{t_2}$	8
11	$\forall yRt_2y$	3	$?_\exists$	10
13	Rt_2t_1	5	$?_{t_1}$	12
15	Rt_2t_2	5	$?_{t_2}$	14

Fig. 3. Play of a material dialogue relative to $\mathcal{M}$, won by **O**

In this play all atomic sentences uttered by **O** are true in $\mathcal{M}$. **P** cannot make any further move since there is nothing for him to defend, and he has attacked already twice all utterances of **O**. As a matter of fact, there is a winning strategy for **O** in the material dialogue about $\neg\exists x\forall yRxy$ played relative to $\mathcal{M}$: **P** cannot improve his payoff by making his choice of repetition rank differently.

O		**P**	
		$\exists xPx \rightarrow \forall xQx$	0
1	$n := \omega$	$m := 1$	2
3	$\exists xPx$ 0	$\forall xQx$	4
5	$?_{\mathsf{t}_1}$ 4	$Q\mathsf{t}_1$	6
	$\vdots$	$\vdots$	
$2n + 3$	$?_{\mathsf{t}_n}$ 4	$Q\mathsf{t}_n$	$2n + 4$
	$\vdots$	$\vdots$	

Fig. 4. Play of a material dialogue relative to $\mathcal{N}$, won by **P**

In this play **O** has granted herself the right to repeat any question a countable infinity of times (i.e., as many times as there are individual constants referring

13 A model of a given vocabulary τ may be regarded as a representation of a possible world looked upon from the perspective of limited conceptual resources (those embodied by the constant and relation symbols of τ).

to elements of the domain of $\mathcal{N}$). Repetition does not help **O**, however, since **P** can safely utter $Q\mathtt{t}_i$ for all constants $\mathtt{t}_i$. Observe that while **O** could not attack the thesis of the dialogue without uttering the antecedent of this implication, it would be pointless (though perfectly possible) for **P** to counterattack **O**'s utterance of $\exists x Px$: player **O** could easily defend herself by uttering $P\mathtt{t}_7$, and therefore this counterattack would be useless in the sense that afterwards **P** would in any event need to defend the claim $\forall x Qx$. On the other hand, there is indeed a way for **P** to successfully defend the consequent $\forall x Qx$, as illustrated in Figure 4. Actually it is clear that there is a winning strategy for **P** in the material dialogue about $\exists x Px \rightarrow \forall x Qx$ played relative to $\mathcal{N}$.

For Lorenzen, the task of philosophy applied to language was to methodically reconstruct linguistic practices, which would then permit us to understand how those practices were possible in the first place. The role of philosophy was, then, 'synthetic.' Given that his general meaning theory[14] was based on norms and actions, such reconstructions would take the form of specifying rules giving rise to the relevant linguistic practices. The investigation would proceed from phenomena to their constitutive rules. Now, typically different practices rely on mutually conflicting rules. For a concrete example, insofar as the dialogue rules for classical and intuitionistic formal dialogues can be taken to offer a reconstruction of classical logic and intuitionistic logic, respectively, these logics are based on conflicting sets of rules: one can follow rule (4b) without thereby following rule (4a). For Lorenzen, *preferences* between different practices were a matter of rationality considerations pertaining to their dialogical reconstructions. This might lead one to prefer, say, the intuitionistic dialogues over and above the classical ones (as Lorenzen in fact did), but it does not mean that there *are* no other kinds of practices, reconstructed by different sets of rules. Lorenzen was particularly clear about the possibility of formulating logical dialogue rules in different ways in his first publication on his game-theoretical interpretation of logic, 'Logik und Agon.' There he not only considered classical logic, but went so far as to say (without apparent irony) that it is fortunate that in mathematics we need not abandon classical logic [29, p. 194] — that by modifying the intuitionistic dialogue rules we may obtain another set of rules providing a reconstruction of classical logic[15]. (Later in his career Lorenzen became increasingly hostile to classical logic and classical mathematics and would probably not have pronounced about the prospect of retaining mathematics based on classical logic in that way.) As to intuitionistic logic, Lorenzen took his intuitionistic dialogue rules as justifying the logical ideas of Brouwer and Heyting, whom he considered to have arrived at the right logic for wrong reasons[16].

[14] When we speak of 'meaning theories' we intend theories about meaning in the generic sense (i.e., what Peacock [38] has proposed to call 'theories of meaning'), though in practice we concentrate on the particular case of first-order logic (and therefore on one specific 'meaning theory' in Peacock's sense).

[15] K. Lorenz attempted in his doctoral thesis *Arithmetik und Logik als Spiele* (1961) to characterize in a game-theoretically precise way the difference between intuitionistic and classical dialogues; see [35, pp. 17–95].

[16] See [30, pp. 193–194], [33, p. 39].

Dialogically reconstructing an independently established logic $\mathcal{L}$ means exhibiting a suitable set of dialogue rules which capture the logic in the sense of allowing to prove the following 'equivalence theorem': for any sentence A of $\mathcal{L}$, A is valid if and only if there is a winning strategy for $\mathbf{P}$ in the formal dialogue $\mathcal{D}(A)$. The dialogicians' claims about formal properties like those expressed by such 'equivalence theorems' have often dramatically outrun what has actually been warranted by their arguments. Many of these failures have been documented by Felscher, who also was the first, in 1985, to substantiate the 'equivalence theorem' for intuitionistic first-order logic — claimed to hold already in Lorenzen's 1958 talk[17]. There is, however, another way of dialogically approaching existing logics. Instead of attempting a reconstruction, one may consider how different existing logics have been motivated, and then proceed to see what such a motivation would mean in the dialogical framework. In this way Shahid Rahman and his collaborators have studied free, relevance, paraconsistent and linear logics [41, 43, 44, 45]. Here the idea is not to find suitable rules so as to dialogically reconstruct, say, relevance logic, but to see what happens when the ideas that have motivated relevance logic are looked upon from the dialogical viewpoint. This approach is fruitful from the dialogue-internal perspective, leading to definitions of new 'argumentative practices.' The resulting dialogically defined logics may very well differ in many respects from the corresponding logics as standardly defined. It is an additional research question to study how, for example, the dialogically defined relevance logic relates, for its metatheoretical properties, to the standard relevance logic. (Research of this kind has not been pursued in the literature.)

Dialogicians think that the particle rules capture a non-trivial portion of the semantics of the logical operators. It has been suggested that keeping the particle rules constant while varying the structural rules, one can substantiate such philosophical ideas as logical pluralism[18]. As they are stated, the particle rules are symmetric in the sense that no matter whether we take X to be $\mathbf{P}$ and Y to be $\mathbf{O}$ or *vice versa*, the rules remain the same. Such symmetry is taken to be pivotal for the intended semantic role of the particle rules. Surely the *schematic* rules themselves must look more or less like they do, as long as we assume, as is done in dialogical logic, that we are dealing with formulas which can be attacked and defended in terms of their subformulas[19]. But why should we accept these specific rules — how are they justified, what are the grounds for accepting them? Can they be *justified* in a symmetric fashion, i.e., in such a way that the justification does not depend on the role distribution considered, $X \mapsto \mathbf{P}, Y \mapsto \mathbf{O}$ or

[17] See Felscher [8]. Dialogicians like Rückert however oppose Felscher's formulation of dialogues, suggesting that his dialogue rules are fine-tuned so as to match certain strategy-level desiderata, whereas the conceptual order should be the reverse and strategy-level properties should emerge from an independently motivated definition of the play level (personal communication). Therefore the dialogicians of the latter kind still lack a justification for the claim that *their* dialogues capture intuitionistic logic!

[18] For logical pluralism, see [3]. For dialogical logic and logical pluralism, see [42, 48].

[19] Cf., e.g., [27, p. 197], [28, pp. 41–42].

$X \mapsto \mathbf{O}, Y \mapsto \mathbf{P}$? Insofar as the symmetry of the rules is semantically crucial, the requirement of symmetry appears to extend to the justification of the rules. If the grounds on the basis of which we may accept such rules are sensitive to the role distribution, these rules as parts of our theoretical framework are, on the whole, asymmetric — even if they could be schematically laid down with no reference to role distributions.

If we consider how the particle rules could be justified, we observe that at least two parameters must be fixed before we can proceed: a role distribution and a semantic attribute associated with the type of dialogue at hand. The sorts of commitments that are induced by the utterances of a player depend on his or her role, and these commitments are relative to a semantic attribute. In particular, there appears to be no hope in symmetrically motivating the rule for implication in connection with the semantic attribute of validity. From the viewpoint of validity, the rule can be motivated for the role distribution $X \mapsto \mathbf{P}, Y \mapsto \mathbf{O}$ as follows: if **P** has *asserted* $A \rightarrow B$ and **O** *grants* A, then **P** is committed to *asserting* B under that assumption. On the other hand, the motivation for the role distribution $X \mapsto \mathbf{O}, Y \mapsto \mathbf{P}$ is totally different: if **O** has *granted* $A \rightarrow B$, then in order for **P** to be able to make use of this concession he must be prepared to *assert* A, and if **O** does not go against that assertion she must then *grant* B as well. Further, if we switch to the semantic attribute of material truth, these justifications no longer work, since granting has no role in connection with material truth. The particle rule for implication in connection with material truth is justified by saying that if X asserts $A \rightarrow B$ and Y asserts A, then X must assert B; in this case the justification does not depend on the role distribution but does depend on the semantic attribute. If dialogicians can be forced to accept these allegations, their whole meaning theory will appear circular — because then particle rules require for their justification some specific notion of semantic attribute, which the dialogician however would like to define on the strategic level utilizing the dialogues whose conceptual building blocks those very same particle rules are.

How could the dialogicians meet the above critique? One possibility would be to deny that particle rules are in need of justification to begin with, and to claim that it makes as little sense to demand that they be further justified as it would make to ask for a justification of why conjunction has the truth table it has in classical logic, or why the rules of intuitionistic natural deduction look like they do[20]. To this we could object that at least it should be possible to *explicate* such basic rules in terms of the primitive informal notions of one's theory. For example, proof-conditional semanticists attempt to clarify their inference rules using what they call 'meaning explanations' — using informally understood basic notions of proof and constructive procedure to explicate specific formal proof rules. In their case the informal notion explaining, and the formal notion to be explained, are notions of the same level, so to say. By contrast, if one is to give such an informal explanation to the particle rules — and if the above analysis is correct — then one ends up applying an informally understood notion of semantic attribute to

20 Helge Rückert holds this view (personal communication).

clarify the particle rules. This would be unsatisfactory, as it would suggest that conceptually the particle rules call for their explication the higher-order notion of semantic attribute, after all. As to other possible responses, from Lorenzen's own viewpoint one might refer to his 'methodic' or reconstructive approach and suggest that indeed we start from a situation where we already have notions such as that of semantic attribute available, the whole point being to explicate how the situation emerged. In such a synthetic approach there is no pretense that we could jump outside the linguistic practice in the midst of which we find ourselves. The goal is to understand the practice while partaking to it. Some reference to higher-level notions in explicating their own emergence might be warranted. Anyway, insisting that we just accept the particle rules in their schematic form — requiring no further justification or explication — appears too cheap a trick to be conceptually convincing.

3 GTS

Like Lorenzen in particular and the dialogical approach in general, also Hintikka sees a crucial connection between meanings of expressions and correlated actions. But whereas for the dialogicians the relevant actions are utterances of sentences, for Hintikka the actions are related to words and phrases, which can then be used to compose sentences. Also, the actions in question are *not* any sort of speech acts (e.g., they do not consist of uttering a word). Instead, in cases crucial to creating semantic relationships between language and reality — 'reality' standing here for what the language is about and what the language is used to deal with — the activities are nonverbal and they consist of linking words and phrases to objects in the world. In important cases (quantifiers) the objects to which logical words are related are non-linguistic. The activities are governed by rules. Practices maintaining such rules give rise to 'language games' which Hintikka terms *semantic games*. Semantic relations do not prevail abstractly and independently of human practices; to the contrary, they exist in and through semantic games. Semantic games *constitute* semantic relations[21]. The rules giving rise to semantic games are not argumentative or discursive, as we proposed to label the dialogicians' rules; they are *correlative* or *projective*. A move in a semantic game consists of creating a correlation between a word and an object; it is not any sort of communicative act.

Semantic games are two-player games defined for a language fragment. Here we restrict attention to (classical) first-order logic. The task, then, is to single out rules which correlate suitable actions with quantifiers and connectives, providing meanings of these expressions. Assuming a pre-existing mastery of *symbol meaning* — proficiency with the meanings of the items in the non-logical vocabulary (constant and relation symbols) — the role of the GTS rules is to associate logical words with activities which give them their meaning[22]. When such rules are

[21] For semantic games and Hintikka's interpretation of Wittgenstein's language games, cf. e.g. [12, *Ch.* 3], [16, pp. 40–41, 67–69], [18, pp. 190–194, 212–236].

[22] In GTS, *interpreted* first-order languages are considered. For 'symbol meaning' and 'sentence meaning' in connection with GTS, cf. [14, p. 26].

provided for all logical words of the language fragment in question, they jointly define, for every sentence of that fragment, a semantic game. More specifically, when properly formulated, they determine a set of partial plays (sequences of moves made in accordance with the game rules), as well as conditions for winning and losing a play. Now, it is a trivial but consequential general feature of games that as soon as the winning conditions and the set of partial plays of a game are fixed, also all strategic properties of the game are fixed[23]. Unavoidably we get two things at the price of one. The basis formed by the game rules gives rise to the superstructure on which notions making use of strategic considerations reside. In connection with semantic games there appear at the level of the superstructure the notions of truth and falsity (in a model). Hintikka [14, pp. 127–128] speaks of the *'two hats' problem* when referring to the initially perhaps perplexing fact that semantic games can serve two purposes which are as different as giving meaning to the logical operators and constituting the notion of truth/providing sentence meanings. He takes *sentence meaning* to consist of the condition that must be satisfied in order for a sentence to be true. Failing to understand the two sides of all games — play level and strategy level — could lead someone to think that insofar as we can speak of truth at all, we need special game rules for the word 'true,' so that any attempt to apply the notion of truth in connection with one language game would force us to move to another language game. Hintikka maintains that when the notion of game is correctly understood, it is seen that the notion of truth is constituted by the very same language games that yield the meanings of logical operators — it is neither a metalogical notion nor a matter of switching to a fresh language game. Turning to speak of the truth of first-order sentences simply marks a step from considerations pertaining to particular moves to considerations pertaining to winning strategies of a suitable player [14, p. 128].

Speaking of first-order logic, what are, then, the activities that according to Hintikka's analysis are associated with quantifiers? As Hintikka explains [16, pp. 41, 67], Wittgenstein had taken it as a criterion for something to be an object that it can be looked for and found. Applying this idea to quantifiers with such objects as values, Hintikka took the activities associated with quantifiers to be looking for and finding[24]. What does it mean, more specifically, that the meanings of the quantifiers are given by means of these activities? Think of the institution of searching and (possibly) finding. It can be viewed from two perspectives. On the one hand, we may think of searching as an activity which is carried out by simple acts such as spotting an object and observing whether it is of the sort looked for. One can perform such acts blindly, without any discrimination as to which object to inspect. Carrying out activities like searching admits of repetitions. Getting disappointed once or twice means nothing: the search can go on. Typically, then, searching lends itself to a whole panorama of realizations, some leading to success (finding), others not. This brings us to the

[23] In technical terms, the extensive form of a game uniquely determines its strategic form.

[24] For a critique of this idea, see [24] and cf. [37, pp. 7–9].

second perspective on the institution of looking for and finding: the strategic perspective. Since among the multitude of acts by means of which to implement the task of searching, there are good ones and bad ones (relative to the purpose, viz. finding), and since in general it depends on the circumstances which acts are to be preferred to which others, searching presents itself as a strategic task — in addition to being, from the first-mentioned perspective, merely a simple matter of blindly performing an act. Conceptually the blind acts are what first and foremost constitute the language–world links and what other people can observe us executing, without subjecting our actions to their efforts of understanding. Any attempts to see us acting in a goal-oriented fashion or to analyze our actions in terms of notions such as intentions are results of a leap to the strategic level. In brief, there are two ways to understand the activities of searching and finding, to which we may refer as the 'blind interpretation' and the 'strategic interpretation.' The former gives rise to the latter: indicating all possible blind acts and specifying which of those yield the desired outcome, is all that is needed for answering all questions of the strategic nature. This notwithstanding, when speaking of the activities of looking for and finding, there is a tension between the two interpretations and it is crucial to keep them apart. Hintikka [14, p. 128] discusses basically the same distinction by speaking of 'definitory rules' and 'strategic rules' of semantic games — the former laying down winning conditions as well as the available moves at a given position, the latter incorporating comparisons between moves in terms of better and worse. Speaking of *rules* in both cases may involve some risk of confusion. In order to specify a game, the only rules needed are the rules of the former kind; the latter 'rules' are just systematizations of strategic facts, themselves entirely fixed by the definitory rules.

A further feature of institutions such as that of searching and finding is that they may be considered as involving interaction of two players, call them player 1 and player 2[25]. To remain on the heuristic level for a while, think of player 2 as the one who searches, while player 1 can affect the circumstances in which player 2 may perform the act of selecting an object for inspection. In cases where searching must result in success (finding) in a number of different circumstances, not just in one particular isolated case, there must be a strategy yielding a suitable act for player 2 depending on the circumstances chosen by player 1. We may, then, say that player 1 co-searches and seeks to co-find[26]. Also the activities of player 1 can be seen from two perspectives (play level, strategy level). But crucially, the success of player 1 is measured differently: what is bad for the other is good for him. So if the result of the co-search by player 1 is a certain scenario in which player 2 picks out an object *not* of the suitable sort, player 1 wins. Else if player 2 hits on a suitable object in that scenario, player 1 loses. As to the strategic side, player 1 has a way of guaranteeing a personally advantageous outcome if he can single out a scenario in

[25] Let us agree that player 2 is female ('she') and player 1 male ('he').

[26] Hintikka does not employ such terminology. Here the prefix 'co-' is intended to connote duality, not anything done jointly.

which player 2 *cannot* find a suitable object, no matter to which object the latter turns attention. Player 2, again, holds the keys to an outcome advantageous to her if for any scenario resulting from the co-search of the adversary, she can find a suitable object. When the activities of looking for and finding are considered in the context formed by a web of further activities, or simply when these activities themselves are iterated, the 'circumstances' or 'scenarios' in which player 2 must make her choices grow strategically more involved — while the relevant blind actions remain of the same kind, totally irrespective of the context.

Let us consider the semantic games for first-order logic in some detail. These games are played relative to a model $\mathcal{M}$ of some non-logical vocabulary τ. Once such a model is fixed, it is also fixed which atomic sentences, producible in vocabulary τ, are true and which are false[27]. That a truth-value distribution among the relevant atomic sentences is given at the outset is a special case of the more general assumption that the relevant *symbol meanings* be given. (Their being given means that the interpretations of the non-logical symbols are fixed separately for every model in some appropriate class of models.) In addition to the two players of the semantic games, there are two *roles* to consider: call them $\mathbb{V}$ and $\mathbb{F}$[28]. Bijective maps $\rho : \{\mathbb{V}, \mathbb{F}\} \to \{1, 2\}$ are *role distributions.* The transposition of a role distribution ρ is denoted by ρ^* and satisfies $\rho^*(\mathbb{V}) = \rho(\mathbb{F})$ and $\rho^*(\mathbb{F}) = \rho(\mathbb{V})$. If $\mathcal{M}$ is a model of vocabulary τ, a is an element of its domain, and $\mathsf{c} \notin \tau$, then we write $\mathcal{M}^{\mathsf{c}}_a$ for the model which is otherwise like $\mathcal{M}$ but interprets the constant symbol c as the element a[29]. For every first-order sentence A, model $\mathcal{M}$ and role distribution $\rho : \{\mathbb{V}, \mathbb{F}\} \to \{1, 2\}$, a semantic game $G(A, \mathcal{M}, \rho)$ is associated:[30]

1. Suppose $A = R(\mathsf{c}_1, \ldots, \mathsf{c}_n)$. If $\mathcal{M} \models A$, the player whose role is $\mathbb{V}$ wins while the player whose role is $\mathbb{F}$ loses; otherwise $\rho(\mathbb{F})$ wins and $\rho(\mathbb{V})$ loses.
2. If $A = \neg B$, then the play continues as $G(B, \mathcal{M}, \rho^*)$. That is, the players switch roles and the play continues with the sentence B.
3. If $A = (B \wedge C)$, then $\rho(\mathbb{F})$ makes a choice between *left* and *right*, and if D is the corresponding conjunct, the play continues as $G(D, \mathcal{M}, \rho)$.
4. If $A = (B \vee C)$, then $\rho(\mathbb{V})$ makes a choice between *left* and *right*, and if D is the corresponding disjunct, the play continues as $G(D, \mathcal{M}, \rho)$.
5. If $A = \forall x B(x)$, then $\rho(\mathbb{F})$ picks out some element a of the domain of $\mathcal{M}$, gives it a name unless it already has one, and if c is the name, the play continues as $G(B[x/\mathsf{c}], \mathcal{M}^{\mathsf{c}}_a, \rho)$. In other words, the player having the role $\mathbb{F}$ chooses an object out there and baptizes it if needed, whereafter the play continues with the result of substituting the relevant constant symbol c for

---

[27] If α is *atomic*, then '$\mathcal{M} \models \alpha$' abbreviates '$\alpha$ is true in $\mathcal{M}$.'

[28] '$\mathbb{V}$' stands for *verifier* and '$\mathbb{F}$' for *falsifier*. For how to understand these terms, see Section 5.

[29] The model $\mathcal{M}^{\mathsf{c}}_a$ is an *expansion* of $\mathcal{M}$ to the vocabulary $\tau \cup \{\mathsf{c}\}$.

[30] In GTS, implication is not given a game rule of its own; $(A \to B)$ is taken to be defined as $(\neg A \vee B)$.

all free occurrences of x in $B(x)$. If the element had initially no name, then as a result of the baptizing we no longer consider the model $\mathcal{M}$ but its expansion, having an interpretation even for the new symbol c.

6. If $A = \exists x B(x)$, then $\rho(\mathbb{V})$ picks out some element a of the domain of $\mathcal{M}$, gives it a name unless it already has one, and if c is the name, the play continues as $G(B[x/\mathsf{c}], \mathcal{M}_a^{\mathsf{c}}, \rho)$. Here it is the player whose role is $\mathbb{V}$ who chooses an element and names it if needed.

Recall the two examples of plays of material dialogues depicted in Figures 3 and 4; here are examples of plays of the corresponding semantic games.

Example (a): Let us start with the sentence $\neg\exists x \forall y Rxy$ and the model $\mathcal{M}$ whose domain consists of two objects which already have a name, t_1 and t_2. Initially player 1 has the role $\mathbb{F}$ and player 2 the role $\mathbb{V}$. The first thing that happens when the game is played is that the players switch roles: player 1 assumes the role $\mathbb{V}$ and player 2 the role $\mathbb{F}$, and the play continues with the unnegated sentence $\exists x \forall y Rxy$. Then the player whose role is $\mathbb{V}$, viz. player 1, chooses an element of the domain. Suppose he picks out t_2. Thereafter the play continues with respect to $\forall y R\mathsf{t}_2 y$, and the player whose role is $\mathbb{F}$, i.e., player 2, must choose an element. Suppose she chooses t_1. So the play has come to an end. Since the atomic sentence $R\mathsf{t}_2\mathsf{t}_1$ is true in $\mathcal{M}$, the player whose role is $\mathbb{V}$ — namely player 1 — wins the play. Observe that here there is no way of continuing the play once the atomic level has been attained. In the corresponding material dialogue this is possible, provided that player **P** has not yet consumed the repetition rank he had initially chosen. As a matter of fact, there is a winning strategy for player 1 in the semantic game correlated with the sentence $\neg\exists x \forall y Rxy$ and the model $\mathcal{M}$. It consists of choosing the element t_2 for $\exists x$. Both possible choices of player 2 for $\forall y$ lead to a win for player 1.

Example (b): Let us look at the sentence $\exists x Px \rightarrow \forall x Qx$ and the model $\mathcal{N}$. The domain of $\mathcal{N}$ consists of objects carrying the names t_1, t_2, t_3, etc. First recall (from footnote 30) that in connection with GTS implication is taken to be defined in terms of disjunction and negation: $\exists x Px \rightarrow \forall x Qx$ abbreviates $\neg\exists x Px \vee \forall x Qx$. Since initially player 1 has the role $\mathbb{F}$ and player 2 the role $\mathbb{V}$, it is player 2 who starts by choosing a disjunct. Suppose she picks out the right disjunct. The the play goes on with the sentence $\forall x Qx$ and the player whose role is $\mathbb{F}$ selects an element of the domain, say t_{127}. The play has come to an end with the atomic sentence $Q\mathsf{t}_{127}$. Since this sentence is true in $\mathcal{N}$, the player whose role is $\mathbb{V}$ — i.e., player 2 — wins. Note that while the play of the corresponding material dialogue depicted in Figure 4 is infinitely long, the play of the relevant semantic game just described involves only two moves. Actually, there is a winning strategy for player 2 in the semantic game associated with the sentence $\neg\exists x Px \vee \forall x Qx$ and the model $\mathcal{N}$; it consists simply of choosing the right disjunct.

In GTS one might wish to *define* the model-relative notions of truth and falsity of a first-order sentence using semantic games in the following way, letting ρ_0 or the 'initial role distribution' satisfy $\rho_0(\mathbb{V}) = 2$ and $\rho_0(\mathbb{F}) = 1$[31].

- *Truth:* A is true in $\mathcal{M}$, if there is a winning strategy for player 2 in $G(A, \mathcal{M}, \rho_0)$
- *Falsity:* A is false in $\mathcal{M}$, if there is a winning strategy for player 1 in $G(A, \mathcal{M}, \rho_0)$.

The GTS definition of what it means for a sentence to be true is linked to the standard model-theoretical Tarski-type definition by the following fact. Assume the Axiom of Choice. Then for any model $\mathcal{M}$ of any vocabulary τ, and for any first-order sentence A written in the vocabulary τ, we have: $\mathcal{M} \models A$ holds in the Tarskian sense if and only if there is a winning strategy for player 2 in game $G(A, \mathcal{M}, \rho_0)$[32].

By inspecting the game rules it is immediate that for any first-order sentence A, all plays of game $G(A, \mathcal{M}, \rho)$ come to an end in finitely many steps. The number of steps needed is bounded above by the maximum number of nested logical operators in A. This may be compared with the case of dialogues (both formal and material), where the syntax of the thesis does *not* alone suffice to yield a finite bound to the length of a play: this is why the players choose the repetition ranks. The reader may check, for instance, that taking the sentence $\neg\forall x \exists y R(x, y)$ as the thesis of a formal dialogue, whether classical or intuitionistic, player **P** can by his choice of repetition rank make a play of the dialogue exceed any given finite length, while player **O** forces a win simply by passively responding to **P**'s questions and waiting until **P** could only continue by overrunning his rank. For comparison, any play of the semantic game associated with the sentence $\neg\forall x \exists y R(x, y)$ — played relative to any suitable model and with either role distribution — is made up of three steps, one of which transposes the roles and two of which are moves consisting of the players' picking out an object — player $\rho(\mathbb{F})$ first, player $\rho(\mathbb{V})$ then. When comparing GTS to other model-theoretical approaches to semantics, it is worth noting that in GTS we need neither to introduce variable assignments nor to assume that all elements of the domain carry a preassigned name; the former assumption is needed in the objectual Tarski-type semantics of quantifiers and the latter in the substitutional Tarski-type semantics. In particular GTS does not interpret quantifiers substitutionally; they are evaluated by choosing an extralinguistic object from the domain (which is then given a name unless it already has one), cf. [12, pp. 102–104]. Quantifiers are processed by choosing locally — relative to a play — names to stand as signposts for the objects picked out. When a new play is considered,

---

31 Semantic games for first-order logic are games of perfect information with *finite horizon* (for each game there is a fixed finite upper bound to the number of moves in a play of that game). By the Gale-Stewart theorem, then, semantic games are determined.

32 See [20, pp. 6–7], [23, p. 94]. Hodges [24] notes that we can avoid the assumption of the Axiom of Choice by adopting a weakened notion of strategy (nondeterministic strategies).

those temporary labels are deleted; they are not kept in store for future usage. Hintikka and Kulas [20, pp. 87–112] have suggested that linguistic phenomena such as anaphora might be best analyzed with reference to such play-relative labels standing for individuals.

As was already stressed, moves made in accordance with GTS rules have nothing to do with speech acts such as uttering. In fact, the players of the semantic games must *not* be thought of as actual language users or as participants of a conversation. Hintikka has attributed to Peirce the contrary view according to which the players of language games are the utterer and the audience, cf. [17, p. 538]. In Hintikka's own view, language users do *not* play semantic games. Instead, they reason about semantic games. They make hypotheses and predictions about them. Asserting a sentence involving quantifiers is to make a claim about what can and what cannot happen when a certain language game is played. Using language involving quantifiers requires mastering the rules of the corresponding semantic games[33]. They provide the relevant semantic background for language use. Hintikka has ascribed this latter view on language games to Wittgenstein.

4 Threats of Circularity

Both dialogical logic and GTS are meant by their exponents to offer a substantial meaning theory for logical words. In this respect they differ from Tarski's model-theoretical truth definition, which cannot offer a substantial meaning theory: its precondition is that logical words such as 'and,' 'not,' and 'all' are already understood, and it uses them to recursively bestow truth conditions on a recursively defined set of strings[34]. It would take a gross misunderstanding of the goal of a model-theoretical truth definition for strings involving symbols such as $\wedge$, $\neg$ and $\forall$ to think that it attempts to offer a substantial meaning theory for logical words.

It was suggested above that in a closer analysis dialogical logic does not fare well in avoiding circularity. Dialogue rules relate utterances to other utterances. Particle rules are supposed to lay down the commitments to which an utterance leads, as well as to explicate the ways of testing such commitments. Whether the dialogician says that one is committed to the uttered sentence's having a certain semantic attribute, or to the defensibility of the utterance against any critique by the adversary, the notion of semantic attribute is presupposed: the attribute of there existing a certain sort of strategy. Commitment is a strategic notion, in an explosive manner: not only does commitment require finding a response to all attacks; any one of those responses gives rise to a new commitment. We are caught in circularity: while the goal is to capture a semantic attribute at the strategic level, it is presupposed already at the play level — not in the sense that particle rules could not be laid down with no reference to strategic notions, but in the sense that the justification of those particle rules appears unavoidably to require ascent to the strategic level. What could be a way out for the dialogician?

[33] See, e.g., [12, pp. 63–66], [14, p. 128], [16, p. 538].
[34] Cf., e.g., [12, p. 58, footnote 12].

As already suggested in Section 2, one option would be to insist that particle rules are in need of no justification precisely because they are, according to the dialogican, the primitive rules which are responsible for the meanings of logical operators. Another option is to actually present a credible account of particle rules which completely avoids resorting to strategic notions. And a third option would be to say that particle rules are a part of the dialogical reconstruction of the linguistic reality in the midst of which we find ourselves, and for this reason they may be motivated in terms of characteristics (such as semantic attributes) that are in an unreconstructed form available to be theorized about, even though in the reconstruction they emerge only on the basis of those particle rules. None of these options appears to be developed in the literature in a detailed and satisfactory manner. If this conceptual problem about game rules cannot be solved, one must give up the idea that dialogues have a meaning-theoretically fundamental role; they are then reduced to mere systematizations of ways of intersubjectively agreeing upon the semantic attribute of a sentence, rather than providing a substantial analysis of meanings of the logical operators and yielding an explication or definition of what it is for a sentence to have a given semantic attribute.

Is GTS, then, definitely beyond the reach of any criticism of circularity? Promisingly for GTS, the linguistic entities on which the game rules operate (logical words) are *not* of the same type as those linguistic entities (sentences) to which the relevant semantic attributes (truth, falsity) are ascribed. The game rules are in terms of actions that pertain either to the world (quantifiers) or to the syntax (junctions) or correspond to a role shift (negation). Yet, GTS cannot in any obvious way be said to go against Frege's context principle according to which words only have meaning in a sentential context. While the blind activities of looking for and finding, considered in isolation, arguably avoid presupposing the notion of truth, these activities are in GTS employed to provide the meanings of quantifiers (and other logical expressions) precisely relative to sentential contexts — as already a superficial inspection of the GTS rules reveals. In fact, the relevant sentential context serves to define what is looked for, and provides the criteria of having found! E.g., the game rule for the existential quantifier tells us that if the position is $\exists x B(x)$, the player whose role is $\mathbb{V}$ picks out an object, call it c, whereafter the play goes on from the position $B[x/\mathsf{c}]$. The choice by the relevant player is supposed to be the result of a blind search. What was looked for? An object satisfying the formula $B(x)$. On which condition is c a successful choice? On the condition that the sentence $B[x/\mathsf{c}]$ is true. Hintikka [14, p. 26] wishes to reassure us that the notion of truth is not involved in his game rules, but is this view tenable? Technically speaking we *can* recursively generate the set of all partial plays of a semantic game — just as we can, for that matter, generate all partial plays of a dialogue — without presupposing semantic attributes. The problem is interpretational: what justifies the chosen rules? In the case of GTS, how do the recursive rules manage to constitute the meanings of quantifiers without presupposing the notion of truth? The problem appears genuine since the rules are relative to a sentential context and the objects looked for are described in terms of the truth of sentences obtained syntactically from that very sentential context. There is a way out, but it makes the GTS rules appear semantically less innocent than Hintikka seems willing to admit. The justification of the game rules is not circular: (**a**) in order to motivate the rule applied to the quantifier Qx in a sentence $QxB(x)$ of quantifier rank $n+1$, the associated activities of searching and finding can be explicated with reference to the truth of sentences of quantifier rank at most n. What is also crucial is that (**b**) the act of processing the sentence $QxB(x)$ does not even implicitly involve the attribution of truth to *this* sentence (or any other sentence of quantifier rank at least $n+1$), it is all about relating the logical word Qx to an object of the domain, relative to a sentential context which is specified by a sentence of quantifier rank at most n. It *would* be a misrepresentation of Hintikka's ideas to suggest that his game rule, say, for the existential quantifier incorporates the following blatantly circular justification: in order for the sentence $\exists xB(x)$ to be *true*, there must be a witness somewhere, call it c, so that $B[x/\mathsf{c}]$ is *true*. Such a transition from $\exists xB(x)$ to $B[x/\mathsf{c}]$ is of course truth preserving, but as a justification for Hintikka's game rule it would be circular, since truth is a strategy-level notion and can only on pain of circularity be applied to the input of a game rule when the justification of that game rule is being considered. The dialogicians' approach defies a clarification like the one just presented in items (*a*) and (*b*) for GTS: in order to justify the particle rules, we must explicate the *ascription of a semantic attribute to an utterance* in terms of ascriptions of semantic attributes to logically simpler utterances. Even if the GTS rules are not circular in this way, the notion of truth seems to force its appearance to the explication of GTS rules, rendering their justification a more subtle matter than first meets the eye. In the present paper we content ourselves to pointing out that this issue deserves attention when discussing merits and dismerits of GTS as a meaning-theoretical approach.

5 Ambiguity of 'Verification'

The way in which GTS defines the notion of truth of a sentence makes it not so far-fetched to think of GTS as some sort of 'verificationist' meaning theory. Winning strategies of player 2 in the semantic game $G(A, \mathcal{M}, \rho_0)$ 'verify' the sentence A relative to the model $\mathcal{M}$. These strategies, then, may look like perfect candidates for means of establishing the truth of A in $\mathcal{M}$. Such a verificationist perspective may seem to link semantic games in an interesting way to proof-theoretical considerations and to the anti-realist or justificationist meaning theory of philosophers like Dummett and Prawitz,[35] as well as to meaning-theoretical ideas underlying dialogical logic. It is the purpose of this section to clarify in which sense these impressions are correct and in which sense they are not.

[35] For work in this direction, see [47]. For problems, see [14, pp. 22–45], [19, pp. 34–35, 38–42]. By 'anti-realism' I mean the idea according to which truth-ascriptions are meaningful only in the presence of means of recognizing whether the ascription is correct (cf., e.g., [7, Ch. 5]).

It must be noted, to begin with, that in spite of its 'verificationist' character, the meaning theory to which GTS gives rise is perfectly truth-conditional. Truth conditions originate from semantic games. The condition for sentence A to be true in model $\mathcal{M}$ is simply that there exists a winning strategy for player 2 in game $G(A, \mathcal{M}, \rho_0)$. The notion of verification emerging from GTS must be kept apart from the notion of verification operative in versions of anti-realism. Hintikka's position is *not* one of an anti-realist. Actually, as Hintikka [13] argues, GTS offers a middle ground between entirely 'static' variants of the truth-conditional view and the verificationist views which lay stress on the epistemic capacities of the language users.

From the anti-realist perspective, it makes sense to ascribe truth only to a sentence for which we possess a means of recognizing it as true; similar remarks apply to attributions of any semantic attribute, say falsity or validity. The dialogical meaning theory could arguably be seen as a variant of anti-realism: ascribing a semantic attribute to the sentence A amounts to there being a means for **P** to force, in a suitable dialogue about A, intersubjective recognition of A's having this attribute. At the very least, winning conditions for individual plays of dialogues have such an anti-realist flavor (cf. footnote 11). For the realist, by contrast, meaningful sentences have truth conditions which prevail or fail to prevail independently of our epistemic efforts such as inference or argumentation; it is one thing for a sentence to be true and quite another for us to be in a position to say that we know it is true. The realist considers it as a fatal mistake to think that the activities used for establishing material truth could be modeled on proofs — activities by means of which we establish logical truths (validities), cf., e.g., [19, pp. 35, 38]. This is because for the realist the proofs of logical truths are simply a matter of symbol manipulation in accordance with rules whose correctness does not depend on what the world happens to be like; their application is a language-internal activity and as such cannot possibly create the sorts of links between language and the world that are the *sine qua non* of all semantics. The anti-realist, again, finds it entirely appropriate to liken processes of establishing material truths to logical and mathematical proofs, since all of these are means of gaining knowledge of semantic attributes of sentences. The anti-realist would, furthermore, not accept the allegation that these activities remain on the level of symbol manipulation; these activities are rather seen as constitutive of sentence meaning.

For purposes of disambiguation, let us use the terms 'verification$_1$' and 'verification$_2$' so that the latter stands for winning strategies of player 2 in semantic games, whereas the former stands for real-life verification processes[36]. More specifically, 'verification$_1$' will be used to designate notions such as means

[36] The distinction *verification$_1$/verification$_2$* resembles the distinction *proof/proof object* made in Martin-Löf's type theory (cf., e.g., [51]). There, proofs are what provide epistemic access to proof objects. On the other hand, the latter distinction has its most obvious domain of application in connection with formal proofs, and therefore proof objects are to be compared primarily with **P**'s winning strategies in formal dialogues rather than with winning strategies of player 2 in semantic games.

of gaining knowledge (of the truth of a sentence), or recognizing the truth of a sentence, or establishing the sentence as true. Also processes or strategies or plans or 'methods' of knowledge-seeking are counted as verifications$_1$. In a generalized sense mathematical proofs, as well as proofs establishing logical truths (validities) are verifications$_1$. Let us take an example. Consider evaluating the sentence 'There are diamonds in Kuhmo' relative to the actual world. The following could be the specification of a real-life procedure for coming to know the truth of this sentence — or putting oneself in a position to ascribe truth to it: drive 120 kilometers to the northeast, follow the marked path, proceed to drill, send the sample to the laboratory and wait for the results of the analysis. By contrast, a winning strategy of player 2 in the relevant semantic game would consist of singling out coordinates for a diamond lying in a kimberlite pipe within the boundaries of Kuhmo. As this example for its part illustrates, there *are* unmistakable formal similarities between verifications$_1$ and verifications$_2$, but there are also crucial differences. Verifications$_1$ deliver *knowledge* of truths, they are means of knowledge acquisition — instead of merely constituting the requisites for a sentence's being true. Verifications$_2$, again, are those strategy-level entities emerging from semantic games whose existence constitutes the truth of a sentence. Their precise relationship is that verifications$_1$ implement verifications$_2$ or are their epistemically accessible realizations. Verification$_1$ of the sentence A aims to produce verification$_2$ of this sentence in a way that allows us to recognize that a verification$_2$ of A has been yielded — thereby guaranteeing not only that A is true, but also that we know it to be true; cf. [14, p. 35].

Regarding the two notions of verification, it may be noted that while in logical empiricism certain sorts of verifications$_1$ were used to formulate a criterion of *meaningfulness* of sentences, anti-realism is more liberal in meaningfulness attributions but uses verifications$_1$ to articulate a criterion for ascribing *truth* to sentences. In GTS, verifications$_2$ constitute truth. Only when attention is turned from truth to knowledge of truths, verifications$_1$ enter the picture as implementations of verifications$_2$. Semantic relations exist only via the activities which give rise to semantic games, but it does not follow that the question of the truth of a sentence depends on the epistemic abilities of the language users, cf. [14, pp. 42–43]. The meaning theory of GTS is *not* a variant of anti-realism. Just because the semantic links are created by human activities does not mean that there are no objective facts about those links. Attempts to gain knowledge of such facts may be blurred by epistemic shortcomings, but this does not imply that the facts themselves are not objective.

An important part of the reason why the views of Hintikka and the anti-realists are so totally at odds with each other on the issue of truth lies in the sorts of activities that the two parties take to provide meanings of linguistic expressions. The anti-realists consider actions pertaining to sentences. They conclude that the assertibility conditions of a sentence such as $\forall x B(x)$, if it involves quantification over an infinite totality, can never be satisfied, since we cannot possess means of recognizing the requisite infinity of facts[37]. In GTS, again, the actions are related to words used to compose sentences, and it is a perfectly manageable human activity to associate the quantifier $\forall x$ with a single object in an infinite domain. The truth of $\forall x B(x)$ is not a matter of a one-time ascription whose justification is subject to our human limitations; it is constituted by the totality of the different ways in which the quantifier $\forall x$ may be correlated with an object, any such object satisfying the formula $B(x)$. From the viewpoint of GTS, the notion of truth emerges via activities giving meanings to words. It might be worthwhile to stress that consequently there is no need for separate language games for 'truth.' Also, we do not learn to apply the notion of truth case by case, depending on the sort of sentence and the sort of circumstances at hand.

6 A Fourfold Division of Foundational Views

Van Heijenoort [10] distinguished between two essentially different ways of understanding the scope and limits of logic and the character of logical theorizing: logic as *language* and logic as *calculus*. The former view (which van Heijenoort attributes to Frege) is characterized by the idea of the *universality* of logic. We cannot meaningfully consider varying the domain of quantification; there is one and only one universe, it is fixed, and it comprises all that there is. Further, we cannot assume an outsider's perspective on logic; we cannot theorize about it from above, metatheoretical questions are not meaningful, and we cannot systematically vary interpretations of non-logical symbols in the sense of developing a model theory or an explicit formal semantics. Positively, logical research takes the form of studying formal systems: we must content ourselves with symbol manipulation — developing explicit semantic rules would require a disinterested perspective that we cannot attain. Under the contrasting view, logic is a calculus in the sense of something that can be *re-interpreted*: the domain of quantification can be changed and interpretations of non-logical symbols systematically varied. Metatheoretical questions can be posed. Van Heijenoort mentions Löwenheim as an early representative of the latter view. The idea of logic as calculus, then, gives rise to model-theoretical considerations.

Hintikka has proposed to generalize van Heijenoort's distinction from logic to language at large, speaking of 'language as the universal medium' and 'language as calculus.'[38] The former view is associated with the 'universalist tradition,' according to which we cannot step outside our language and theorize about it from outside. Semantics in general and the notion of truth in particular are considered ineffable. Varying interpretations and reasoning in terms of alternative 'possible worlds' is not judged meaningful. By contrast, the idea of 'language as calculus' goes together with the 'model-theoretical tradition' within which it is considered possible to explicitly theorize about semantics, vary interpretations and resort to many-world conceptualizations. Hintikka includes Frege, Russell and Wittgenstein

[37] If a concrete model is needed, consider the model $\mathcal{M} = (D, B^{\mathcal{M}})$ with $D = \{a_1, a_2, \ldots\} = B^{\mathcal{M}}$, where a_i and a_j are distinct for any distinct i, j. Thus, nothing else is assumed of the structure of the elements of the domain except that they all satisfy the unary predicate B.

[38] See the articles collected in [15].

in the first-mentioned tradition, and Gödel and Peirce in the second tradition[39]. He refers to the two views as 'ultimate presuppositions' [15, p. 21] because typically they are not expressly asserted by the philosophers adhering to them. Besides, the different ideas lumped together in terms of the distinction need not be neatly distributed in the case of a particular thinker; one and the same person may entertain ideas from both sides. In the 1930s Carnap belonged to the universalist camp, switching to the model-theoretical side during his later period. Tarski, again, applied model-theoretical methods freely to formal languages but was a universalist regarding natural language [15, p. 108].

Hintikka sees the two views as involving contrasting assumptions about the relation between language and reality. The way in which he sees the distinction is predicated on his unquestioned conviction that semantics is a matter of creating and maintaining objectively prevailing language–world links. For the universalist such links are a precondition of our language of which nothing can be said in the language itself, while a representative of the calculus view finds it possible to theorize about such links. On the other hand, we have noted that philosophers like Dummett, Lorenzen and Prawitz adopt a different approach to meaning — an approach where sentences are conferred meaning via epistemically conditioned activities associated with those sentences. It seems preferable to phrase the distinction so as not to rule out at the outset such an alternative meaning-theoretical view. Consequently, let us distinguish between 'universalism' and 'anti-universalism.' Universalism regards it as impossible to explicitly theorize about semantics (no matter how the meanings of linguistic expressions are thought to arise). According to anti-universalism, again, explicit semantic theorizing is possible and we can systematically vary those factors that give rise to meanings of linguistic expressions (again irrespective of one's theoretical standpoint on semantics).

Let us now consider the fourfold division determined jointly by the two distinctions *realism/anti-realism* and *universalism/anti-universalism*. For each of the resulting four categories, we may ask for examples of philosophers belonging to the category. Frege and Wittgenstein of the *Tractatus* are clear examples of realist universalists, and Hintikka is a realist anti-universalist. It appears correct to classify Dummett as an anti-realist universalist[40]. But who could be an

[39] While in particular Frege's distinction between *Bedeutung* and *Sinn* certainly serves to increase awareness of the fact that the referents of lexical items might be different from what they in fact are, only when phrased by Carnap as the distinction between extension and intension in the context of a totality of possible states of affairs did this distinction lead to properly model-theoretical considerations.

[40] Dummett is prepared to consider different domains of quantification and recognizes the possibility of varying interpretations. Yet he thinks that a meaning theory of a specific language deals with a single interpretation: the 'intended' or 'correct' one [6, p. 20]; he maintains that meaning is a one-world issue. Further, proof-theoretically motivated explications of meanings of logical constants and justifications of logical laws are in keeping with the universalist attitude. They are language-internal comments on how aspects of the same language work, not any metasystematic characterizations.

anti-realist anti-universalist? That would mean imposing epistemic criteria on truth ascriptions, but at the same time seeing semantic theorizing as essentially relying on systematic variation of meanings conferred to linguistic expressions. Lorenzen appears to fulfill these criteria. It is correct that he comes close to the universalists in maintaining that we cannot posit ourselves outside our language — that natural language is a boat from which we cannot disembark [32, p. 28] — but he refuses to accept as a consequence that we could not attempt to 'methodically construct' the practical preconditions of our language. The starting point for our theorizing is a complex linguistic practice in the midst of which we find ourselves; the philosophical task is to understand how it has emerged by effecting its 'practical transcendental deduction' from a network of postulated simple actions. While a coherent universalist finds it literally impossible to pronounce upon semantic relations (any comments on them take the form of indirect clues or perhaps metaphors), Lorenzen takes understanding a linguistic practice to consist of postulating suitable simple practices and methodically showing how the complex practice may emerge from these ingredients. Lorenzen's anti-universalism has two roots. On the one hand, a given linguistic practice might admit of a variety of rational reconstructions. On the other hand, one and the same syntax can in principle be associated with a variety of practices; thereby even meanings associated with *logical expressions* can be varied (whereas the model-theoretical variant of anti-universalism varies domains of quantification and interpretations of *non-logical* symbols). The latter possibility is realized notably in the difference between classical and intuitionistic dialogue rules; more specifically, in this case the particle rules remain the same and certain structural rules are varied.

As the considerations in the present paper for their part show, systematically comparing the two game-based approaches to philosophical meaning theory — GTS and dialogical logic — offers insights into many highly interesting issues in the philosophy of logic. In particular, despite the fact that these approaches share some important conceptual tools, the associated philosophical views are in most respects very different indeed.

References

[1] Abramsky, S., Jagadeesan, R., Malacaria, P.: Games and Full Abstraction for PCF. Information and Computation 163(2), 409–470 (2000)

[2] Auxier, R.E., Hahn, L.E. (eds.): The Philosophy of Jaakko Hintikka. Library of Living Philosophers, vol. 30. Open Court, Chicago (2006)

[3] Beall, J.C., Restall, G.: Logical Pluralism. Oxford University Press, Oxford (2006)

[4] van Benthem, J.: Logic, Rational Agency, and Intelligent Interaction. ILLC Prepublication Series, PP-2008-6 (2008)

[5] Clark, R.: Games, Quantification and Discourse Structure. In: [36], pp. 139–150 (2009)

[6] Dummett, M.: The Logical Basis of Metaphysics. Harvard University Press, Cambridge (1991)

[7] Dummett, M.: Thought and Reality. Clarendon Press, Oxford (2006)

[8] Felscher, W.: Dialogues, Strategies, and Intuitionistic Provability. Annals of Pure and Applied Logic 28, 217–254 (1985)
[9] Girard, J.-Y.: Locus Solum. Mathematical Structures in Computer Science 11, 301–506 (2001)
[10] van Heijenoort, J.: Logic as Calculus and Logic as Language. Synthese 17, 324–330 (1967)
[11] Hintikka, J.: Language-Games for Quantifiers. In: Rescher, N. (ed.) Studies in Logical Theory, pp. 46–72. Basil Blackwell, Oxford (1968)
[12] Hintikka, J.: Logic, Language-Games, and Information: Kantian Themes in the Philosophy of Logic. Clarendon Press, Oxford (1973)
[13] Hintikka, J.: Game-Theoretical Semantics as a Synthesis of Verificationist and Truth-Conditional Meaning Theories. In: LePore, E. (ed.) New Directions in Semantics, pp. 235–258. Academic Press, London (1987)
[14] Hintikka, J.: The Principles of Mathematics Revisited. Cambridge University Press, Cambridge (1996)
[15] Hintikka, J.: Lingua Universalis vs. Calculus Ratiocinator: An Ultimate Presupposition of Twentieth-Century Philosophy. Jaakko Hintikka: Selected Papers, vol. 2. Kluwer, Dordrecht (1997)
[16] Hintikka, J.: Intellectual Autobiography. In: [2], pp. 3–84 (2006)
[17] Hintikka, J.: Reply to Wilfrid Hodges. In: [2], pp. 535–540 (2006)
[18] Hintikka, J., Hintikka, M.B.: Investigating Wittgenstein. Basil Blackwell, Oxford (1986)
[19] Hintikka, J., Kulas, J.: The Game of Language: Studies in Game-Theoretical Semantics and Its Applications. Reidel, Dordrecht (1983)
[20] Hintikka, J., Kulas, J.: Anaphora and Definite Descriptions: Two Applications of Game-Theoretical Semantics. Reidel, Dordrecht (1985)
[21] Hintikka, J., Sandu, G.: On the Methodology of Linguistics: A Case Study. Basic Blackwell, Oxford (1991)
[22] Hintikka, J., Sandu, G.: Game-Theoretical Semantics. In: van Benthem, J., ter Meulen, A. (eds.) Handbook of Logic and Language, pp. 361–410. Elsevier, Amsterdam (1997)
[23] Hodges, W.: Elementary Predicate Logic. In: Gabbay, D.M., Guenthner, F. (eds.) Handbook of Philosophical Logic, vol. 1, pp. 1–131. Reidel, Dordrecht (1983)
[24] Hodges, W.: Logic and Games. In: Zalta, E. (ed.) The Stanford Encyclopedia of Philosophy (2006) (Summer 2006 Edition), `http://plato.stanford.edu/archives/sum2006/entries/logic-games/`
[25] Hyland, J.M.E., Ong, L.: On Full Abstraction for PCF I, II, and III. Information and Computation 163(1), 285–408 (2000)
[26] Japaridze, G.: Introduction to Computability Logic. Annals of Pure and Applied Logic 123, 1–99 (2003)
[27] Kamlah, W., Lorenzen, P.: Logische Propädeutik oder Vorschule des vernünftigen Redens. Bibliographisches Institut, Mannheim (1967)
[28] Lorenz, K.: Dialogspiele als semantische Grundlage von Logikkalkülen. Archiv für mathematische Logik und Grundlagenforschung 11, 32–55, 73–100 (1967)
[29] Lorenzen, P.: Logik und Agon. In: Atti del, X.I.I. (ed.) Congresso Internazionale di Filosofia (Venezia, 1958), pp. 187–194. Sansoni, Firenze (1960)
[30] Lorenzen, P.: Ein dialogisches Konstruktivitätskriterium. In: Infinitistic Methods, pp. 193–200. Pergamon Press, New York (1961)
[31] Lorenzen, P.: Metamathematik. Bibliographisches Institut, Mannheim (1962)
[32] Lorenzen, P.: Methodisches Denken. Suhrkamp, Frankfurt am Main (1968)

[33] Lorenzen, P.: Normative Logic and Ethics. Bibliographisches Institut, Mannheim (1969)
[34] Lorenzen, P.: Die dialogische Begründung von Logikkalkülen. In: Gethmann, C.F. (ed.) Theorie des wissenschaftlichen Argumentierens, pp. 43–69. Suhrkamp, Frankfurt am Main (1980)
[35] Lorenzen, P., Lorenz, K.: Dialogische Logik. Wissenschaftliche Buchgesellschaft, Darmstadt (1978)
[36] Majer, O., Pietarinen, A.-V., Tulenheimo, T. (eds.): Games: Unifying Logic, Language, and Philosophy. Springer, Heidelberg (2009)
[37] Marion, M.: Why Play Logical Games? In: [36], pp. 3–26 (2009)
[38] Peacock, C.: The Theory of Meaning in Analytical Philosophy. In: Fløistad, G. (ed.) Contemporary Philosophy, pp. 35–56. Nijhoff, The Hague (1981)
[39] Pietarinen, A.-V.: Most Even Budged Yet: Some Cases for Game-Theoretic Semantics in Natural Language. Theoretical Linguistics 27, 20–54 (2001)
[40] Prawitz, D.: Meaning and Proofs: On the Conflict Between Classical and Intuitionistic Logic. Theoria 43, 2–40 (1977)
[41] Rahman, S., Carnielli, W.: The Dialogical Approach to Paraconsistency. Synthese 125, 201–232 (2000)
[42] Rahman, S., Keiff, L.: On How to Be a Dialogician. In: Vanderveken, D. (ed.) Logic, Thought and Action, pp. 359–408. Springer, New York (2005)
[43] Rahman, S., Rückert, H.: Dialogische Logik und Relevanz. Universität des Saarlandes, memo no. 27 (December 1998)
[44] Rahman, S., Rückert, H.: Eine neue dialogische Semantik für lineare Logik. In: [49], pp. 141–172 (2007)
[45] Rahman, S., Rückert, H., Fischmann, M.: On Dialogues and Ontology. The Dialogical Approach to Free Logic. Logique et Analyse 160, 357–374 (1997)
[46] Rahman, S., Tulenheimo, T.: From Games to Dialogues and Back: Towards a General Frame for Validity. In: [36], pp. 153–208 (2009)
[47] Ranta, A.: Propositions as Games as Types. Synthese 76, 377–395 (1988)
[48] Rückert, H.: Why Dialogical Logic? In: Wansing, H. (ed.) Essays on Non-Classical Logic, pp. 15–30. World Scientific, London (2001); reprinted in [49], pp. 15–30
[49] Rückert, H.: Dialogues as a Dynamic Framework for Logic. Ph.D. thesis. Department of Philosophy, Universiteit Leiden (2007)
[50] Saarinen, E.: Dialogue Semantics versus Game-Theoretical Semantics. In: Proceedings of the Biennial Meeting of the Philosophy of Science Association, pp. 41–59 (1978)
[51] Sundholm, G.: Proof-Theoretical Semantics and Fregean Identity Criteria for Propositions. The Monist 77, 294–314 (1994)
[52] Tulenheimo, T.: Comparative Remarks on Dialogical Logic and Game-Theoretical Semantics. In: Bour, P.E., Rebuschi, M., Rollet, L. (eds.) Construction. Festschrift for Gerhard Heinzmann, pp. 417–430. College Publications, London (2010)

Natural-Language Syntax as Procedures for Interpretation: The Dynamics of Ellipsis Construal

Ruth Kempson[1], Eleni Gregoromichelaki[1], Wilfried Meyer-Viol[1], Matthew Purver[2], Graham White[2], and Ronnie Cann[3]

[1] Philosophy Department, King's College London
[2] School of Electronic Engineering and Computer Science, Queen Mary University of London
[3] Linguistics and English Language, University of Edinburgh

Abstract. In this paper we set out the preliminaries needed for a formal theory of *context*, relative to a linguistic framework in which natural-language syntax is defined as procedures for context-dependent interpretation. Dynamic Syntax provides a formalism where both representations of content and context are defined dynamically and structurally, with time-linear monotonic growth across sequences of partial trees as the core structure-inducing notion. The primary data involve *elliptical fragments*, as these provide less familiar evidence of the requisite concept of context than anaphora, but equally central. As part of our sketch of the framework, we show how apparent anomalies for a time-linear basis for interpretation can be straightforwardly characterised once we adopt a new perspective on *syntax* as the dynamics of transitions between parse-states. We then take this as the basis for providing an integrated account of ellipsis construal. And, as a bonus, we will show how this intrinsically dynamic perspective extends in a seamless way to dialogue exchanges with free shifting of role between speaking and hearing (*split-utterances*). We shall argue that what is required to explain such dialogue phenomena is for contexts, as representations of content, to include not merely partial structures but also the sequence of actions that led to such structures.

1 Preliminaries

Despite extensive research on the context-dependence of natural-language (NL) understanding over the last thirty years, with formal modelling of a wide range of individual phenomena, there has been little attempt to bring everything together in order to seek an overall concept of context-dependence. In particular, ellipsis has been treated wholly differently from *anaphora*, despite the fact that, like anaphora, ellipsis is a phenomenon which, by definition, exhibits radical context-dependency. An elliptical construction is one "that lacks an element that is, nevertheless, recoverable or inferable from the context" (Wikipedia: *ellipsis*). This characterisation corresponds to the robust folk intuition that, in

Lecomte and S. Trонçon (Eds.): PRELUDE 2010, LNAI 6505, pp. 114–133, 2011.

language use, expressions can be omitted because context fully determines the way the fragment utterance is to be understood. Seen from this point of view, it is reasonable to expect that the phenomenon of ellipsis will provide a basis from which the notion of context needed for language interpretation can be explored; and indeed in this paper we shall use investigation of ellipsis exactly to this end. As [Purver et al., 2006; Cann et al., 2007] argue, if suitably dynamic concepts of structure and context are defined, a unitary account of ellipsis, in all contexts, can be provided making sense of all the distinct aspects of context-dependence exhibited in language use.

2 Ellipsis and the Syntax-Semantics Interface

Current accounts of ellipsis do not in general purport to provide a point of departure for the study of context [though cf. Ginzburg and Cooper, 2004; Fernández, 2006]. The consensus is that ellipsis is not a homogeneous phenomenon. Rather, it splits into syntactic, semantic and pragmatic types, with only the last type depending on context for construal.

The general background for both syntactic and semantic accounts is the methodology of conventional grammars which dictates the sentence as the unit of characterisation: the only forms of ellipsis addressed have been those where the ellipsis site can in some sense be analysed sententially – either as a second conjunct of a compound conjunctive form or as an answer to a question with both of these being analysed in sentential terms:

(1) A: Have you seen Mary?
B: Mary? No, I haven't. I have, Bill. Tom too.
(a) (b) (c) (d)
B: Have I seen Mary? No, I haven't seen Mary but I have seen Bill; and I have seen Tom too.

Thus (1a) can be understood as an echo of the original question, (1b) as the negative answer *I have not seen Mary*, and so on. Indeed (1) illustrates a number of different ellipsis types. Each of them has been argued to be a separate syntactic phenomenon on the evidence of apparently different structural constraints governing their reconstruction as full sentential forms. However, the ever-accumulating set of phenomena, all labelled ellipsis, do not seem reducible to some general abstract account. Indeed, the general phenomenon, ellipsis, remains puzzling, with apparent conflicting evidence for both semantic and syntactic forms of explanation [see Ginzburg and Cooper, 2004; Merchant, 2007]. On the one hand, a semantic explanation is available for cases where, for a single antecedent form and assigned interpretation, ambiguity nonetheless arises:

(2) John checked over his mistakes, and so did Bill/Bill too.
'Bill checked Bill's mistakes' ("sloppy")
'Bill checked John's mistakes' ("strict")

This is argued to reflect a process of abstraction over some content provided by the antecedent ('John checked over John's mistakes') creating distinct abstracts to apply to the content of the fragment in the elliptical conjunct (a process involving *higher-order unification*): (a) abstraction over solely the subject of the antecedent, or (b) abstracting over the subject and all other references to the individual it denotes. Two resulting predicates can then apply to the elliptical fragment which yields so-called strict/sloppy ambiguities [Dalrymple et al., 1991]. Since these abstraction operations affect content (not syntactic structure) this provides the semantic basis for explanation. A second argument for semantic accounts of ellipsis is provided with the almost invariant parallelism of the mode of interpretation between an elliptical second conjunct and its antecedent clause, as in e.g. parallelism of quantifier dependencies (scope):

(3) Every professor got to meet with a visiting government official, and so did every administrator.

Such parallelism of interpretation is said to arise in virtue of variation in the abstraction steps available for the antecedent clause at the ellipsis site. If abstraction applies *prior* to quantifying in the quantified expression within the predicate, then, with only a variable in place of this quantified form in that first conjunct, the result will be an interpretation in which the quantifier simultaneously binds a variable in both conjuncts, hence incorporating the entire conjunction within its scope. If, however, the abstraction operation to create the requisite predicate takes place *after* all quantified expressions in the first conjunct are quantified in, then whatever quantifying terms are contained in the resulting abstract will be interpreted as taking narrow scope with respect to the subject expression with which the created predicate is to be combined. So far so good. But there are three possible interpretations for a sentence such as (3), not just two. The third is where the indefinite is construed as taking wider scope than the subject NP of the conjunct in which it is contained but nevertheless not wide scope with respect to the whole conjunction. In (3) this involves there being two visiting government officials, one visiting the professors, and one visiting the administrators. This third interpretation cannot be captured by the Dalrymple et al. account (an observation due to Mark Steedman), so the semantic account is at best incomplete.

In any case, there is competing evidence that sensitivity to structure is essential to the way in which the elliptical fragment is reconstructed. Ellipsis is highly sensitive to syntactic/morphological requirements set up by the sequence of expressions preceding the ellipsis site:

(4) A: Who did every husband visit? B: His wife.

Moreover, there are cases of ellipsis, so-called *antecedent-contained ellipsis*, that display sensitivity to the very constraints taken to be diagnostic of syntactic phenomena. These are the so-called 'island' constraints which are taken as evidence that at least those forms of ellipsis must be analysed in syntactic terms [Fiengo and May, 1994; Merchant, 2004]:

(5) John interviewed every student who Bill already had.

(6) *John interviewed every student who Bill ignored the teacher who already had.

In (6), an instance of *antecedent-contained ellipsis*, the ellipsis site cannot be associated with the relative pronoun (*who*) even though this is possible in (5). This is because a relative-clause boundary intervenes between the relative pronoun and the elliptical element. This type of violation is directly redolent of the island restrictions constraining long-distance dependencies as in e.g. *wh*-questions (*wh* binding also not being possible across a relative-clause boundary). Because such restrictions are taken not to be expressible in semantic terms – the lambda calculus underpinning semantic combinatorics would impose no such structure-particular restriction – they have been taken as evidence for a level of syntactic structure independent of semantics, and a diagnostic of what constitutes a syntactic phenomenon. Hence, so the argument goes, at least some types of ellipsis require syntactic explanation, involving full projection of clausal structure at the ellipsis site with subsequent deletion of phonological material. However, even though structural restrictions can be captured by syntactic reconstructions of ellipsis, this provides no explanation of parallelism effects, which require the definition of independent constraints [see e.g. Fox, 1999].

Over and above the division of ellipsis into semantic and syntactic types, there is the very widespread use of fragments in dialogue. Even though grammatical formalisms have neglected these as "performance data", more recently, as dialogue modelling has developed, this omission is being repaired. Extending the Dalrymple et al. pattern, Ginzburg and Sag [2000], Ginzburg and Cooper [2004] defined multiple types of abstraction mechanisms to reflect distinct identified types of ellipsis, ranging over semantic, syntactic and morphological specifications associated with the antecedent of the elliptical form. But the move made is not straightforward, for, retaining the sentence-based methodology, these fragments are analysed as full sentences, with type-lifting of the fragment not merely in the semantics but also in the syntax. This assumption is problematic if the fragment occurs, as in (7), at a point in the dialogue exchange when no antecedent of appropriate type is available in context to yield the appropriate abstract:

(7) A: And er they X-rayed me, and took a urine sample,
took a blood sample.
A: Er, the doctor
B: Chorlton?
A: Chorlton, mhm, he examined me, erm, he, he said now they
were on about a slide [unclear] on my heart. [BNC: KPY 1005-1008]

An additional problem for any such assumption, is that anaphoric and quantificational dependencies can be seamlessly continued across from one speaker to another, as in (4), involving the binding of a pronoun by a quantifying expression. Any syntactic dependency whatsoever can be split between speaker and hearer, with the hearer-turned-speaker continuing an initiated utterance relative to the structural context provided by the first part (see Purver et al. [2009]):

(8) A: Have you read
B: any of your books? Certainly not.

(9) A: Are you OK? Did you burn
B: myself? Fortunately not.

Beyond the syntax/semantics controversy, there are yet further cases where there is no linguistic basis for assigning interpretation to the fragment. Stainton [2006] argues that such cases do not allow any analysis as sentential reconstructions but have to be seen as a speech act that is performed without recourse to a sentential structure:

(10) A (coming out of lift): McWhirter's?
B: Second left.

In these, it is non-linguistic aspects of context that determine the way the fragment is understood. What this illustrates is that, from a pre-theoretical point of view, fragments in general can occur whenever the context provides elements relative to which the fragment can be processed in an appropriate way. The context may provide linguistic structure on the basis of which the fragment may yield a propositional content as in (4), in the fragment interruptions of the questions in (8)-(9) and their subsequent fragment replies. But, as in the case of anaphora (e.g. *bridging* phenomena, see Clark [1977]), sometimes both contents and the appropriate context for processing have to be constructible on the fly as in (10). Even further, the fragment expression may have to be interpreted as an extension of a nonpropositional structure given in the context, as in (7). The disparate ways in which elliptical fragments can be understood have been taken as evidence of the so-called "fractal heterogeneity" of ellipsis [Ginzburg and Cooper, 2004] in that its resolution apparently involves cross-cutting constraints across whatever information the grammar manipulates, with morphological, syntactic, semantic, even phonological information yielding multiple bases for ellipsis. In addition, the grammar interacts with a range of dialogue interpretation principles which also contribute to disambiguating the function of each type of fragment. In the complexity that results from this reconstruction of context as multiple sets of constraints on interpretation across distinct modes of representation, the robustness of the folk intuition is getting lost: how can the context, from which speakers and hearers freely draw, require such complex cross-module constraints and heterogeneity in need of disambiguation online? The challenge of providing a uniform account of ellipsis processing thus remains.

An alternative is to take ellipsis as a phenomenon from which we can glean evidence of the types of information that context records, and on that basis, to explore processes of dynamic update manipulating this information in a unified way. And for this, we turn to Dynamic Syntax, where the dynamics of how information accrues along the time-line of processing is integral to the formalism's structural underpinnings.

3 Dynamic Syntax

Dynamic Syntax (DS) is a model of how interpretation is built up relative to context, reflecting how hearers (and speakers) construct interpretations for strings of words incrementally using linguistic information and context as it becomes available. Crucially, the output of any processing task is a representation of the content of the string uttered in a particular context (not a representation of some hierarchical structure defined over the string, i.e., not a sentence *type* assumed to hold over any linguistic context). NL syntax, on this view, is conceived of as the *process* by which such semantic representations are built up, using instructions associated with words and contextual information to drive the incremental development of the output representation. Since pragmatics may interact at any point with the online syntactic process, output semantic representations may differ for the same string uttered in different contexts.

Formally, DS is a lexicalized grammar using labelled sequences and trees as basic data structures. The labelled sequences are time-linear sequences of words with accompanying phonological, morphological and word-boundary specification; the trees formalize the semantic, functor-argument structure induced from utterances of such sequences in the form of (unordered) trees labeled by terms of a typed lambda calculus and other process-control labels. This emphasis on strings and trees DS shares with other tree-based linguistic theories like Transformational Generative Grammar (in all its guises), Tree Adjoining Grammar, Head-driven Phrase Structure Grammar, and Categorial Grammar. It differs from virtually all these in the way it relates the two data structures. Other theories identify linearity with the yield of ordered trees; in DS, the string determines a sequential 'derivation tree'; the words are interpreted as instruction packages to construct parts of a lambda term in a labelled tree representation. The packages are executed word-by-word from-left-to-right. The terminal nodes of a lambda term produced by a grammatical sentence in some instances stand in one-to-one correspondence with the words of the sentence, but an individual word may also induce sub-structure containing more than one labelled node; and there is no direct relation between the yield of the eventual tree and the sequential order of the string. Terms of only a small, fixed, set of semantic types are used, so no new types (functions) can be constructed (unlike categorial frameworks). Structural underdetermination and update replace concepts of 'movement' (and its analogues in non-transformational frameworks) and function-composition.

The tree is incrementally constructed by tree substitution and addition or extension of labels, but there are two ways to escape the strict linear evaluation order: by *structural underdetermination* and by *underdetermination of labels*; both may lead to delay in choices to be made. Structural underdetermination involves the addition of a sub-term to the tree, an *unfixed node*, whose location in that tree is characterised as merely dominated by a previously constructed term without, as yet, a fully specified hierarchical position. So, for example, in the characterisation of *Mary, John likes* instructions license the introduction of a typed sub-term into the tree as specified by the word *Mary* that cannot yet be given a fixed location in the tree skeleton of the term under construction.

Underspecification in the labels dimension is implemented by means of employing *meta-variables*, indicated as **U**, **V**, etc., as temporary labels allowing identification from some larger context given their associated type specification (and any other constraints that may be identified, such as gender, person, and so on).

The parsing process is lexically driven and directed by *requirements*. The main requirement is '$?t$', the *type requirement* to produce a formula of proposition type t from the word string; but imposition of such requirements for other types induces the creation of new labeled tree structure, providing the basis for expressing how one word may 'subcategorize' for others. As there are only a finite number of types, type requirements can be viewed as *non-terminal symbols* to be rewritten to a terminal symbol (supplied by a word). Tree-growth *actions* are defined as general or lexically triggered options, dividing into two broad types: those for developing a tree-structure for which a number of strategies may be available; and those annotating non-terminal nodes through algorithmic application of β-reduction and its associated type-deduction. These in combination yield a resulting structure, with all nodes properly annotated and no requirements outstanding. Type requirements are not the only form of requirement. Indeed all forms of underspecification are associated with requirements for update, for example, meta-variables co-occur with a requirement for instantiation, represented as $?\exists\mathbf{x}.Fo(\mathbf{x})$. Well-formedness for a string is defined in terms of the possibility of constructing a proper tree rooted in type t with no outstanding requirements left on any node in the tree.

3.1 Tree Growth and LOFT: The Logic of Finite Trees

To round out this summative characterisation, we now sketch the general process of parsing as the induction of a sequence of partial trees, whose input is a one-node tree annotated with only the requirement $?t$ and a pointer $\Diamond$ indicating the node under development, and whose output is a binary branching tree whose nodes reflect the content of some propositional formula. The left-to-right parse of the string *Bob upset Mary* gives rise to the sequence of partial trees shown in Fig. 1, with the final tree completing the initial requirement.

The parsing task, using both lexical input and information from context, is thus to progressively enrich the input tree to yield a complete output using general tree-growth actions, lexical tree-growth actions, and, when triggered by some lexical item, pragmatic tree-growth actions. As all types of action are defined in the same terms, i.e., as actions that map one partial tree to another, different types of action can interleave at any point. *Decorations* on nodes include content-representing formulae, type specifications, a treenode indicator (with one node distinguished as the rootnode: $Tn(0)$), and requirements for such decorations as imposed by the unfolding process. The primitive types are types e and t as in formal semantics but construed syntactically[1]. Each node must be eventually decorated with a pairing of formula and type specifications written as $\alpha : \phi$ (α the formula labelling ϕ the type). Annotations on non-terminal nodes are induced algorithmically in the final evaluation of the trees, in terms inspired by

[1] There are other types, but the list is highly restricted.

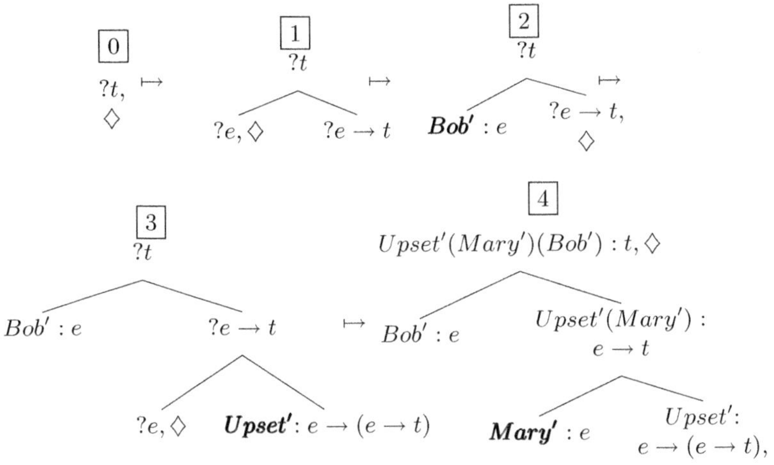

Fig. 1. Monotonic tree growth in DS

the Curry-Howard isomorphism and labelled type-deduction. In all cases, the output is a fully annotated (*decorated*) tree whose topnode is a formula value representing a proposition derived by the string of words processed relative to a particular context of utterance[2].

At the heart of the formal characterisation is the (modal) logic of finite trees [LOFT: Blackburn and Meyer-Viol, 1994] which permits addressing any node in the tree from the perspective of any other node using the immediate-dominance modalities $\langle\downarrow\rangle$ and $\langle\uparrow\rangle$ and variations over these. Such operators can be used to indicate nodes that exist already in the tree (e.g. $\langle\downarrow\rangle\alpha$ indicates that there is a daughter of the current node decorated by label α), with variants distinguishing $\downarrow_0/\downarrow_1$ as argument/functor daughter respectively, and Kleene * operations over these to define general dominance relations. Such decorations are used in conjunction with the requirement operator ?, to indicate nodes that, at some stage, must occur on the tree but may not currently do so (e.g. $?\langle\downarrow\rangle\alpha$ indicates that there must, eventually in the derivation, exist a daughter of the current node decorated by label α). Tree-growth is thus defined relative to the imposition and subsequent satisfaction of requirements: $?X$ for any annotation on a node constitutes a constraint on how the subsequent parsing steps must progress, i.e. X must be derived. Hence requirements such as $?t$, $?e$, $?e \rightarrow t$ impose constraints on tree development that formulae of the relevant types must be provided by the end of the parsing process. Constraints on growth may also be modal, e.g.

[2] We simplify the exposition here: the full presentation of Dynamic Syntax includes the assumption that the proposition expressed is relative to a time point given by a term denoting some temporal/modal relation to the time of utterance [Gregoromichelaki, 2006]. A further simplification is that names and words with conceptual content are assumed to be in one-to-one correspondence with concepts, with no attempt to address the substantial issues in addressing the context-sensitivity of either.

while a decoration $?e$ requires a term to be constructed at the current node, $?\langle\downarrow\rangle e$ requires a daughter node to be so decorated. Although type-requirements are the primary drivers of the syntactic process, in principle, any label may be associated with a requirement. For example, $?\exists\mathbf{x}.Fo(\mathbf{x})$ requires some contentful formula value to decorate a node and is associated with the parsing of pronouns and other anaphoric expressions such as auxiliary verbs in elliptical structures.

An essential ancillary notion to that of tree growth is the concept of *procedure* or *action* for mapping one partial tree to another. These are defined in a language involving such commands as $\texttt{make}(\langle\downarrow\rangle)$, $\texttt{go}(\langle\downarrow\rangle)$, $\texttt{put}(\alpha)$, $\texttt{make}(\langle\downarrow_*\rangle)$, $\langle\texttt{IF}..., \texttt{THEN}..., \texttt{ELSE}...\rangle$ etc. Sets of such actions incorporated in individual packages can be either general *computational rules* or *lexical actions* associated with words contributing content-formulae and other aspects of structure. For example, verbs in English are parsed when the pointer resides at a node decorated with $?e \rightarrow t$ induced by a computational rule. The verb itself contributes not only a concept formula (e.g. $Upset'$) but also creates a new node and moves the pointer to the object node (the transition induced is that of transition 2-3 in Fig. 1)[3]. Parsing of an object will then provide the appropriate formula value for this node. Computational rules can then compositionally determine the combination of those formulae to satisfy the requirements remaining in a strictly bottom-up fashion.

3.2 Semantic Underspecification and Update

As indicated earlier, pronouns, and indeed elements at ellipsis sites like auxiliaries, project temporary, underspecified formula values which are required (through the injunction $?\exists\mathbf{x}.Fo(\mathbf{x})$) to be instantiated at some point during the parsing process. This interim value is given as a metavariable, e.g., **U**, **V**..., which satisfies a type requirement, allowing the parse to continue, but leaves the content of a node open to be satisfied perhaps later in the derivation. The update for such a metavariable is given by selection of a proper value from context. Thus all context-dependent linguistic elements may allow as an option the assignment of a value from the parsing process subsequent to their original processing. Consider such a delayed resolution:

(11) *It* emerged *that John was wrong.*

In (11), the metavariable projected by *it* in string-initial position licenses the further unfolding of the parse process, allowing the parse of the verb *emerged.* The pointer then returns to the semantic subject node and permits the parse of the post-verbal complement clause whose content satisfies the requirement for determinate content, yielding an output formula $Emerge'(Wrong'(John'))$ (ignoring tense), without any final indication that an expletive pronoun appeared in the uttered string.

The same dynamics applies to quantifying terms. It is wellknown that indefinites can take wide scope over any previously introduced term:

[3] Formally: $\langle\texttt{IF}\ ?Ty(e \rightarrow t), \texttt{THEN}\ \texttt{make}(\langle\downarrow_1\rangle), \texttt{go}(\langle\downarrow_1\rangle), \texttt{put}(Fo(Upset'), Ty(e \rightarrow (e \rightarrow t))), go\langle\uparrow_1\rangle, \texttt{make}(\langle\downarrow_0\rangle), \texttt{go}(\langle\downarrow_0\rangle), \texttt{put}(?Ty(e)), \texttt{ELSE Abort}\rangle$.

(12) *Every teacher* confirmed that *two students* had written *a report* on *a famous philosopher*. $\forall < \exists < \exists_2 < \exists; \exists < \exists < \forall < \exists_2; \ldots$

But inversion of scope for other quantifiers is available only if an indefinite precedes:

(13) *A nurse* interviewed *every patient*. (ambiguous)

(14) *Most nurses* interviewed *every patient*. (unambiguous)

Within DS, this is addressed through combining underspecification and the selected logic. All noun phrases, even quantified ones, project terms of type e. There is no type-lifting mechanism reversing functor-argument relationships as in generalised quantifier theory: in all predicate-argument arrays constructed within this framework, argument-hood is non-negotiable. Rather, quantifying expressions are analysed in the manner of arbitrary names of predicate-logic natural deduction, as formulated in the *epsilon calculus* of [Hilbert and Bernays, 1939], a conservative, but more expressive, extension of predicate logic. Introduction of such terms in the language is based on the following equivalence:

$$\exists x.F(x) \equiv F(\epsilon x.F(x))$$

This indicates that an existential statement is equivalent to one in which a *witness* of the truth of the statement can appear as an argument of the statement's predicate. Such a witness appears as an *epsilon term* containing as its *restrictor* the predicate itself and denotes some arbitrary entity satisfying the predicate (if such an entity exists).

Exploiting this equivalence, in DS, such terms are employed because they may carry a record of the context within which they occur inside their restrictor. According to the computational rules defined, an intermediate representation of a sentence such as *A man is waving* will take the form $Wave'(\epsilon, x, Man'(x))$ derived by simple functional application. But this will be eventually algorithmically transformed to a formula where an appropriate epsilon term, abbreviated as a below, appears as the argument of the conjunction of the predicates contributed by the common noun and verb:

$$Man'(a) \wedge Wave'(a)$$

where

$$a = (\epsilon, x, Man'(x) \wedge Wave'(x))$$

The restrictor of this term contains a record of the propositional structure that gave rise to it so that it provides a suitable antecedent for subsequent cases of *E-type anaphora* [see Kempson et al., 2001]. Quantifier scope is not expressed as part of the tree architecture but through scope constraints collected incrementally during the parse process. Multiple quantification of course yields more complex restrictor specifications, but the evaluation algorithm applies to arbitrarily complex combinations of terms and with variation in the connective depending on the type of quantifier involved.

This separation of scope dependencies from the representation structure then allows inverse scope readings for sentences such as (12)-(13) to be treated as a form of context-dependency analogous to anaphoric construal. Scope constraints for indefinites are formulated as partially underspecified, a metavariable indicating availability of choice with respect to source of dependency. This is lexically encoded as $\mathbf{U} < x$ for the variable x associated with the indefinite. This simultaneously allows (a) selection of antecedents as the source of dependency among whatever other terms are already constructed within a domain, and (b) license to delay scope-dependency choice until some other noun phrase has been parsed: it is this license for delay (rather than some unrestricted quantifier storage device) which gives rise to scope inversion as in (12)-(13) but not (14).

3.3 Linking Trees through Shared Terms

For the full array of compound adjunct structures displayed in NL, DS employs a license to build paired, so-called LINK*ed*, trees associated through a LINK modality, $\langle L \rangle$. This device is utilised for allowing incorporation within a tree of information that is to be structurally developed externally to it. Relative clause construal, being one core case of adjunction, involves constructing a LINKed tree bearing a requirement that it contains as a sub-term the formula from the source-node from which the LINK relation is defined:

(15) John, who smokes, left

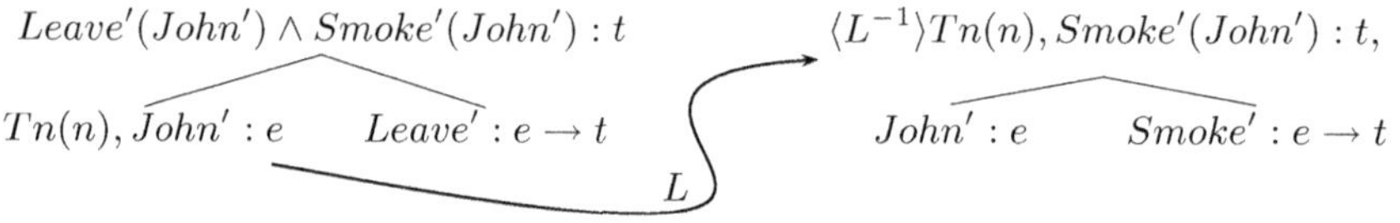

In the building up of such paired LINKed trees for relative clauses, the requirement imposed for a common term is satisfied by the processing of the relative pronoun which induces an unfixed node annotated with the requisite copy of that term. This is then necessarily constrained to appear within the newly emergent propositional structure (reflecting *island constraints* associated with such structures).

A second pattern is provided by *apposition* structures, in which a sequence of NPs are construed as co-denoting:

(16) A friend of mine, a painter, is outside.

(17) The candidate, a linguist, is outside.

In DS terms, the second NP is defined as extending the term derived by processing the first through construction of a LINKed structure. This involves evaluating such paired terms as a single term incorporating both restrictors: $(\epsilon, x, \phi(x))$ and $(\epsilon, x, \psi(x))$ thus leading to the term: $(\epsilon, x, \phi(x) \wedge \psi(x))$, in (16) yielding $(\epsilon, x, Friend'(x) \wedge Painter'(x))$.

Given this dynamic perspective of progressively building up representations of content, we now examine consequent requirements on a model of *context*. As we shall see, given the strictly incremental approach to the processing of strings, context is definable as a composite record of *what has just taken place*, i.e., the representation of (partial) content so far established in tree form and the actions used to gradually build it up[4].

3.4 Intra-sentential Context

We conceive of each parsing step (the processing of each word in a sentence) as taking place in some context. As already set out, a parsing step is defined in terms of tree expansion: its input is a a partial tree including a pointer as built by the preceding steps so far; its output is another, more fully-specified (although still possibly partial) tree. As natural language is inherently ambiguous, a single sequence of words $w_0 \ldots w_i$ might be associated with more than one possible sequence of parsing actions $a_0 \ldots a_i$, $a'_0 \ldots a'_i$, $a''_0 \ldots a''_i$ etc. The parser state when parsing the next word w_{i+1} may therefore contain multiple (partial) trees T_i, T'_i, T''_i etc., according to the degree of ambiguity exhibited by the string of words or the firing of different parsing actions all of which yield well-formed partial trees, i.e. several independent parsing paths may ensue. However, only one such partial tree will provide the basis for a particular update by a particular word. It is thus important that the context in which any lexical action a_{i+1} is applied corresponds only to the partial tree which a_{i+1} is itself extending. If the role of context is to provide information built by previous interpretations, and the interpretations T_i and T'_i are mutually exclusive alternatives, an extension of T_i must not be able to access information from T'_i. Therefore a minimal model of context in DS consists of the current partial tree that is being extended, and ambiguity of parsing paths and/or interpretation may thus be conceived of as involving multiple, but independent, contexts.

As a grammar formalism, nothing internal to DS definitively determines selection of interpretation in cases of genuine ambiguity: in such cases the grammar must make available all the options required. Nonetheless, because of the in-built left-to-right incrementality, a fine-grained concept of *context* is definable incorporating the monotonically growing information provided not only by previous discourse but also by the evolving parse of the current utterance. Such a context provides a record of (a) the partial terms so far constructed, as well as (b) the actions used in constructing them, thus providing the requisite information for resolution of metavariables as well as (as we shall see below) other forms of underspecification.

[4] The formal notion of context in DS also includes the set of words parsed so far but as we will not make use of this here we put it aside for simplicity.

4 Ellipsis and Context

4.1 Inter-utterance Context

When we consider dialogue, our concern with "context" includes what must be provided at speaker-turn boundaries. A new speaker can begin a new sentence, of course, so the default null context is available, in DS terms, the requirement $?t$. Yet a previous complete sentence may provide material enabling a hearer to reconstruct information, as in elliptical utterances, so the context must contain previous (complete) trees. Additionally speakers may continue or collaborate on incomplete utterances spoken by others, as in (7), and this means that we require having incomplete (partial) trees available. We therefore take context after a change of speaker also to contain the current tree (partial or not) from the parser state developed immediately before the change.

For grammar formalisms which define solely the well-formedness of full sentences, this might be problematic, as suitable representations for partial sentences cannot be available without further ado. But with partial trees being well-defined formal objects in the DS representation language, this is straightforward. Moreover, although defined primarily in terms of parsing processes, DS incorporates a parallel account of *generation*. In the DS implementation, generation follows exactly the same steps as parsing, with the added constraint that some desired output tree or *goal* tree (possibly a complete tree as in 4 in Figure 1) constitutes a filter on each transition, so is present from the initial stage (as the content of what the speaker intends to communicate). Hence, parsing/generation processes in DS use interchangeable representations and modelling the distinction between speaking and hearing does not necessitate switching to distinct formal vocabularies or sets of actions [see Purver and Kempson, 2004, as in this paper we set out parsing derivations only].

4.2 Access to Structure

This perspective on context provides all we need to treat common dialogue fragment phenomena such as short answers, acknowledgements and clarification requests: in all these cases, the fragment updates the existing tree representation in the context provided by the preceding utterance. In question/answer pairs as in (18), the tree produced by parsing a *wh*-question includes underspecification (in DS, *wh*-terms encode a specialised **WH** metavariable).

(18) A: Who did John upset?
B: Mary.

If this tree is taken as present in the context in which B's fragment is parsed (19), we can see that parsing the word *Mary* can directly update this tree, through LINK adjunction, as introduced in section 3.3, to the node decorated with the interrogative metavariable, **WH**. The result is to provide a complete semantic formula and an answer to the question via evaluation of the two formula decorations as being equivalent to $Mary'$, which is by definition a legitimate way to further specify metavariables. Such an update is like the appositive further specification of a noun phrase as in *my friend, the doctor* where a pair of noun phrases is taken to identify the same entity, the only difference being that **WH** adds no further information to the term.

(19) Processing *Mary*:

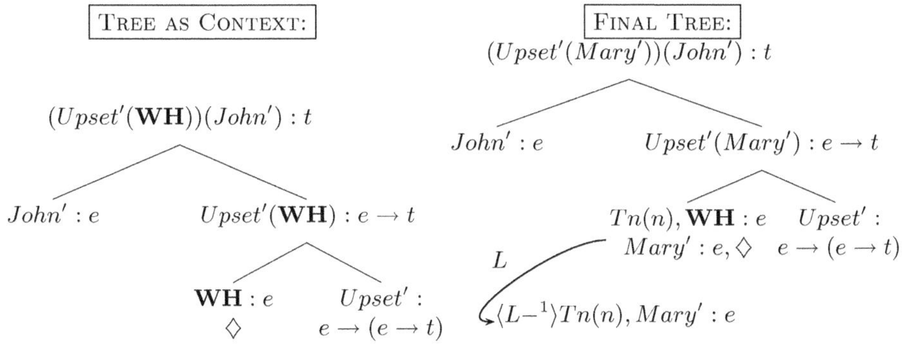

This approach needs no extension for split utterance cases such as (9) in which a reflexive pronoun must be understood relative to what is provided by the first half. Here, the tree in context is partial, as in (20), with the pointer at a node awaiting processing of the object. The actions associated with B's completion of A's utterance as *myself* merely copy the formula at the semantic subject node, just in case that term picks out the current speaker[5]. This is exactly as parsing *yourself* copies the subject formula just in case it picks out the current addressee, a parse of *Did you burn yourself?* thus yielding exactly the same tree as the resulting tree in (20). There is, on this view, nothing mysterious about shifts in perspective within dialogue:

(20) Processing (9):

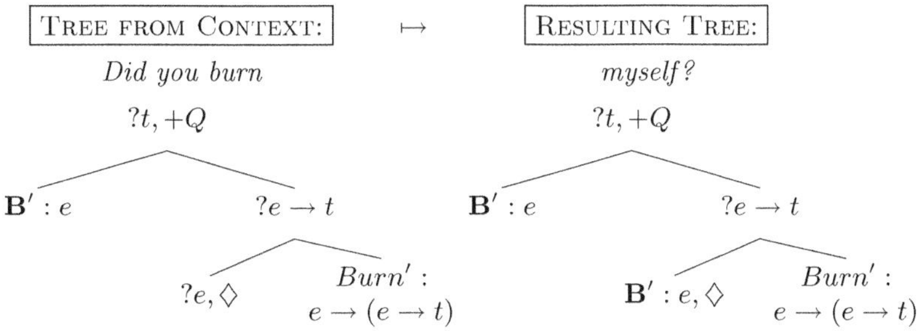

[5] $\langle$`IF` $?Ty(e)$, `THEN` $\langle$`IF` $\langle\uparrow_0\rangle\langle\uparrow_1^*\rangle\langle\downarrow_0\rangle\alpha_{Speaker}$, `THEN` `put`$(\alpha)$, `ELSE Abort`$\rangle$, `ELSE Abort`$\rangle$.

Examples such as (7) also follow, given that the contextual trees at speaker change can show a further degree of (temporary) underspecification: partially specified content. The tree constructed by B after processing the utterance-initial *the doctor* is highly partial, but provides a node of type e which B can use as the basis for a clarification request. Such requests are modelled in DS via the same LINK relations used for appositions. Like appositions, clarifications are interpreted as providing or requesting further information on a specific term. Modelling this via a LINKed sub-tree provides a possible update of the original term, much like an answer to a *wh* question provides an extension of the term queried, as we saw in (18), except that here the clarification question provides additional information: the name of the doctor [see Kempson et al., 2009]:

(21) Processing *Chorlton*:

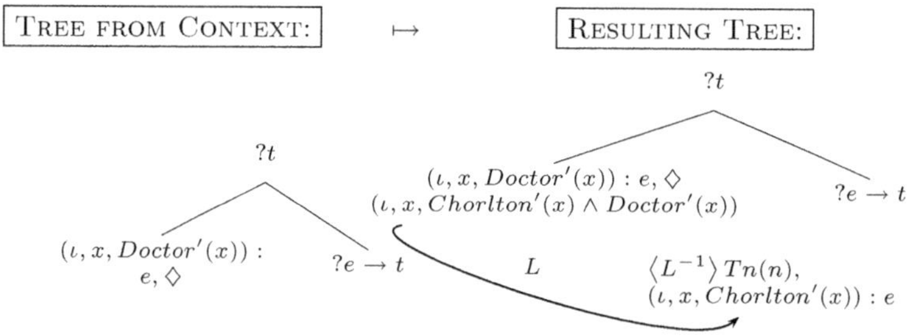

As discussed earlier, switch of roles between the two interlocutors does not impede the term-construction process: it is merely the presence of well-defined partial structures in context, as required by the strictly incremental nature of DS parsing, that makes the analysis possible.

4.3 Access to Formulae

Since context provides us with structure so far built – the labelled, partial trees – it also provides us with the logical formulae decorating these trees. As outlined in section 3.2, this is all we need for the analysis of anaphora: anaphoric elements decorate the tree with metavariables, which require for full resolution the presence in context of a suitably typed formula. The same applies to strict forms of VP-ellipsis, as these can also be modelled as formula underspecification resolved by directly re-using a contextually provided predicate. The only difference between pronoun and ellipsis construal is their logical type assignment, e vs. $(e \rightarrow t)$ (the arrow in the diagram below represents the formal process of *Substitution* of a metavariable by some term in context and **DO** is a specialised metavariable contributed by *did*):

(22) A: Who upset Mary?
B: John did.

Parsing *John did*:

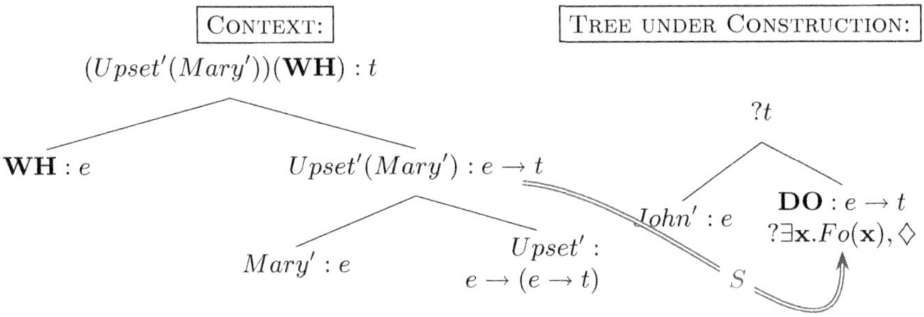

4.4 Access to Parsing Actions

However, cases of *sloppy* ellipsis must require something else: here the processing of the fragment leads to a different interpretation from that of its antecedent – neither an exact copy of the antecedent content nor an extension of it, but still maintaining some parallelism in the way the interpretation is established. We can see now that, given the DS procedural modelling of interpretation, this parallelism can be suitably expressed in terms that do justice to what constitutes context-dependency in such cases: the *actions* used in parsing the antecedent may now be invoked and re-run at the ellipsis site. This provides a new formula value, but one which is constrained to be built in the same way as the antecedent formula was already constructed:

(23) A: Who hurt himself?
B: John did.

Informally, the DS processing for the question in (23) involves the following actions after parsing of the subject *who*: constructing a two-place predicate as indicated by the verb, introducing an object argument, and then, because this object derives from a reflexive pronoun, it is obligatorily identified with the argument provided as subject (for the formulation of such actions see fn. 3,5). When it comes to processing B's elliptical *did*, the reconstruction we require is essentially the same: the same verbal predicate but this time an object which becomes identified with the subject B has provided, i.e., *John*. This can be achieved precisely by re-running the same set of parsing actions as used when parsing A's utterance – provided that these have been stored as a sequence in context and thus are accessible for re-use, in this case the sequence of actions given by *upset+himself*. This is a possibility that, according to DS, is associated with the resolution of the metavariable **DO** contributed by *did*. The effect achieved is the same as the higher-order unification account but without anything beyond what has already been used for the processing of previous linguistic input: the parsing actions themselves stored in context and able to be re-run.

4.5 Context Defined

To sum up, we take as context a record of (a) the partial tree under construction, with its semantic annotations, (b) the trees provided by previous utterances, and

(c) the sequence of parsing actions used to build (a) and (b). Considering the incremental, action-based nature of DS, this reflects well the state of the parser. Trees provide structural context for the incorporation of dialogue fragments; their decorating formulae provide antecedents for anaphora and strict readings of VP-ellipsis; parsing actions allow reconstruction of the sloppy equivalents. A *context*, then, is a sequence of actions and tree representations contributed by a sequence of words and utterances in a discourse.

Examples like (10) might be taken to motivate an extra dimension of context, provided by no obvious linguistic input but instead by the situation or prior knowledge or belief, i.e. the general cognitive context. Nonetheless, the DS semantic representations are expressed in a language intended to implement the interface among information sources in any modality and this kind of example provides confirmation for this strategy. Any type of information is then by definition represented in the same way as the other aspects of context discussed here, namely, as formulae in tree format present within the contextual tree sequence. Such trees would provide the means by which an utterance of *Second left* may be construed as an answer to the elliptical question *McWhirter's?* which itself induced a tree with top node decorated by the formula: $Located'(McWhirter's, \mathbf{WH})$ obtained by exploiting knowledge representations available as context.

There are consequences to be explored from this perspective, both linguistic and formal. With this preliminary characterisation of context, some of the problems facing other accounts of ellipsis become resolvable. Parallelism effects associated with ellipsis such as (3) are expected to follow. The actions used to establish an interpretation for the first conjunct will form part of the immediate context for the interpretation of the second conjunct containing the ellipsis site. So, in the resolution of the ellipsis, these actions will be replicated yielding parallel results in the scopal interpretation of the indefinite. According to DS, the indefinite in (3) can be interpreted with free choice as its being dependent on the subject, the event term associated with the immediately containing proposition, or the event term for the overall formula[6]. Whatever choice is made in the first conjunct will be necessarily replicated in the second as the same actions need to be repeated. This pattern is expected for scopal dependencies of arbitrary complexity, unlike in accounts that involve copying of logical form or strings.

On the other hand, the structural restrictiveness associated with *antecedent contained ellipsis* is predicted as it involves the processing of relative clauses. The relative pronoun in English, given its appearance at the left periphery of its clause, in DS is taken to decorate an unfixed node. Such nodes by definition have to be resolved within the local containing tree, and not across a LINK relation (this is the formal reflection of *island constraints* in DS). This independent constraint is respected in (5) but not in (6). In (6) a second relative clause (*who already had*) intervenes between *who* and its argument position. This induces an island violation and explains the inability of *who* to provide an argument for the predicate to be reconstructed from *interviewed* [Kempson et al., 2001;

[6] DS semantic representations include event arguments (Gregoromichelaki 2006).

Cann et al., 2005]. The island sensitivity of antecedent contained ellipsis in (6) thus follows from this independent constraint on relative clause processing in an entirely expected way.

From this point of view, the notion of context as a record of *previous* parse states remains unchallenged even by the occurrence of expletive pronouns or scope dependencies that are apparently determined non-linearly. In all such cases, the lexical specifications of pronouns/indefinites dictate only partial updates to the node they decorate with metavariables licensing delay – delay of content value assignment for subject expletives and delay of scope-dependency for indefinites. As, in DS, full resolution of underspecified content is not imposed until after the predicate value in some local domain has been determined, the context available for late resolutions of this type will contain terms constructed after the initial, first pass processing of the expletive/indefinite. These terms can then provide values for the underspecified content initially provided. So what superficially appears to be evidence for phenomena that do not conform to the usual context-dependency properties of underspecified expressions reverts to the usual pattern once the concept of context is defined in terms of immediately preceding parse states that can accommodate localised underdetermination.

4.5.1 Formal Considerations

Nonetheless, this novel approach on context specification requires a reconsideration of the DS formal mechanism in order to be expressible in the most apposite terms. DS actions for building up structure currently retain a second-class status, being construction mechanisms which are not themselves quantified over. Several questions on the technical level might then be raised regarding the possibility of a neater formalisation.

One approach might be the following. DS relies on the interrelationship of two different structures: the sequential order of the input (or output) string, and the applicative structure of the tree with typed nodes that it thereby built up (or serialised). We can express both structures using a suitable logical framework based on dependent type theory with subtypes and equality: type dependency can be used to express the sequential order, whereas subtyping can be used to express the applicative structure. To be precise, terms $x : e$, $\phi : e \to t$ and $\phi x : t$ correspond to triples $\langle s, x, \phi \rangle$ inhabiting, respectively, the types $\langle t, e, \{\phi : e \to t | \phi x = s\}\rangle$ – types which are expressible in a suitable dependent type theory and whose dependency corresponds to the serial order of the corresponding elements of the input string. Having done this, further questions present themselves. The first is whether the underlying type theory ought to be linear or intuitionistic: that is, whether it should allow explicit contraction and/or weakening. In fact, both seem to have roles: relative pronouns, for example, seem to behave linearly, whereas anaphora (and E-type phenomena generally) are classical. So we would seem to need a mixed linear/intuitionistic system: the validity of a suitable such system is plausible, but needs further investigation.

In a framework like this, we can pose further questions. One concerns the interaction between type extension and subtyping: systems with dependent intersection – such as that of Kopylov [2003] – assimilate type extension to subtyping, and we have an independent motivation for dependent intersection as a means of expressing the contribution of adjuncts (via LINK-structures). The next concerns the possibility of a constructive variant of the ϵ-calculus, which would allow ϵ-terms which were not everywhere defined (and which would provide a semantic basis for the generation of clarification requests). So we can ask about the compatibility of such a system with ϵ-terms (which includes the question as to how we formulate proof rules for such ϵ-terms). Finally, we can raise the following issue: if, in the spirit of Miller [2009], we formalise algorithms using proof search, can we recover the algorithmic side of DS in terms of a search for the proof of a suitable judgement (a proof, it may be, that the input string inhabits a suitable type)? Such a viewpoint could be a very powerful means of relating DS to other type-theoretic linguistic formalisms.

References

Blackburn, P., Meyer-Viol, W.: Linguistics, logic and finite trees. Logic Journal of the Interest Group of Pure and Applied Logics 2(1), 3–29 (1994)

Cann, R., Kempson, R., Marten, L.: The Dynamics of Language. Elsevier, Oxford (2005)

Cann, R., Kempson, R., Purver, M.: Context and well-formedness: the dynamics of ellipsis. Research on Language and Computation 5(3), 333–358 (2007)

Clark, H.H.: Bridging. In: Thinking: Readings in Cognitive Science, pp. 169–174. Cambridge University Press, Cambridge (1977)

Dalrymple, M., Shieber, S.M., Pereira, F.C.N.: Ellipsis and higher-order unification. Linguistics and Philosophy 14(4), 399–452 (1991)

Fernández, R.: Non-Sentential Utterances in Dialogue: Classification, Resolution and Use. PhD thesis, King's College London, University of London (2006)

Fiengo, R., May, R.: Indices and Identity. MIT Press, Cambridge (1994)

Fox, D.: Economy and Semantic Interpretation. MIT Press, Cambridge (1999)

Ginzburg, J., Cooper, R.: Clarification, ellipsis, and the nature of contextual updates in dialogue. Linguistics and Philosophy 27(3), 297–365 (2004)

Ginzburg, J., Sag, I.: Interrogative Investigations: the Form, Meaning and Use of English Interrogatives. CSLI Lecture Notes, vol. 123. CSLI Publications, Stanford (2000)

Gregoromichelaki, E.: Conditionals: A Dynamic Syntax Account. PhD thesis, King's College London (2006)

Hilbert, D., Bernays, P.: Grundlagen der Mathematik II. Julius Springer, Berlin (1939)

Kempson, R., Gregoromichelaki, E., Sato, Y.: Incrementality, speaker-hearer switching and the disambiguation challenge. In: Proceedings of SRSL 2009, the 2nd Workshop on Semantic Representation of Spoken Language, pp. 74–81. Association for Computational Linguistics, Athens (2009)

Kempson, R., Meyer-Viol, W., Gabbay, D.: Dynamic Syntax: The Flow of Language Understanding. Blackwell, Malden (2001)

Kopylov, A.: Dependent intersection: A new way of defining records in type theory. In: LICS, pp. 86–95. IEEE Computer Society, Los Alamitos (2003)

Merchant, J.: Fragments and ellipsis. Linguistics and Philosophy 27, 661–738 (2004)

Merchant, J.: Three kinds of ellipsis: Syntactic, semantic, pragmatic? Ms University of Chicago (2007)

Miller, D.: Formalizing operational semantic specifications in logic. In: Proceedings of the 17th International Workshop on Functional and (Constraint) Logic Programming (WFLP 2008), vol. 246, pp. 147–165 (2009)
Purver, M., Cann, R., Kempson, R.: Grammars as parsers: Meeting the dialogue challenge. Research on Language and Computation 4(2-3), 289–326 (2006)
Purver, M., Howes, C., Gregoromichelaki, E., Healey, P.: Split utterances in dialogue: a corpus study. In: Proceedings of the 10th SIGDIAL Workshop on Discourse and Dialogue (2009)
Purver, M., Kempson, R.: Incremental context-based generation for dialogue. In: Belz, A., Evans, R., Piwek, P. (eds.) INLG 2004. LNCS (LNAI), vol. 3123, pp. 151–160. Springer, Heidelberg (2004)
Stainton, R.: Words and Thoughts: Subsentences, Ellipsis, and the Philosophy of Language. Oxford University Press, Oxford (2006)

Relevance and Utility in an Argumentative Framework: An Application to the Accommodation of Discourse Topics

Grégoire Winterstein[1] and Gerhard Schaden[2]

[1] Université Paris Diderot-Paris 7
CNRS UMR 7110 LLF
[2] Université Lille 3 Charles de Gaulle
CNRS UMR 8163 STL
{gregoire.winterstein,gerhard.schaden}@linguist.jussieu.fr

In this paper, we address the question of the exact place one should attribute to game theory in the analysis and modelisation of meaning in natural language. One can think of at least three possible positions with respect to this issue:

1. Game theory is not inscribed at all in the grammar, but is an effective framework to describe effects of language use (that is, a formal method of dealing with pragmatics) [1].
2. Every aspect of game theory has grammatical effects, i.e. any kind of pay-off is inscribed in the semantics of linguistic items.
3. Grammar deals with a specific type of games. If this should be the case, linguists need to identify the relevant subpart of game theory.

We will focus in our paper on positions (1) and (3). A plausible candidate for a linguistic phenomenon that can be successfully modeled by a subset of game-theory is *argumentation* (cf. Ducrot (1980)). Argumentation concerns aspects of the meaning of sentences (words) that are not reducible to truth conditions. As such, one may ask whether these aspects are properly grammatical (i.e., encoded in the linguistic system), or pertain to the realm of pragmatics (i.e., contextual effects).

In order to sort these things out, we investigate the link between the notions of *relevance* (in a theory of argumentation, as implemented in Merin's decision-theoretic pragmatics) and the notion of *(expected) utility* in a game-theoretical framework of pragmatics. Crucially, our aim is to see if those two notions can and should be unified, which can only mean that relevance should be reduced to the expected utilities of the discourse participants — which would bring us to position (1), above. We will present arguments against such a move.

To illustrate our point, we will study the accommodation of discourse topics in dialogues. We will show that relevance in an argumentative sense can easily be linked to the establishment of a discourse topic, since both notions are meant to relate to the point of discourse or to what the discourse is all about.

[1] This stance would naturally need to exclude for instance game-theoretic formulations of logics that can be used to represent truth conditions of natural language semantics, as proposed by Hintikka & Sandu (1997).

Lecomte and S. Tronçon (Eds.): PRELUDE 2010, LNAI 6505, pp. 134–146, 2011.

Our hypothesis is that full-fledged game theory, as a very general and powerful theory of the strategic interaction of (rational) agents, should be left to deal with particular tokens of utterances in particular situations. That is, we take it to serve best as a formal theory of pragmatics. Argumentation theory, however, is concerned with general linguistic properties of utterances, that is with the types of utterances, and can yield predictions in the *grammar* of a given language, which in principle is blind to a speaker's utility.

1 Utility and Relevance

1.1 Argumentation and Argumentation Theory

The notion of argumentation in the sens of argumentation theory (as developped by Anscombre and Ducrot, cf. Ducrot (1980), Anscombre & Ducrot (1983)) describes some regularities in natural language that cannot be straightforwardly treated through truth-conditional considerations alone. One classical example of a word with argumentative properties is *almost*. Clearly, *almost* P entails $\neg P$. However, *almost* P argues for the same conclusions as would P, and not for the conclusions that would induce $\neg P$.

In (1-a), even though the first part of the utterance conveys that it is not dark yet, this does not support the targeted conclusion, namely that more than sidelights would be too much. In this respect, (1-a) is identical to (1-b), which has rather different truth conditions. On the other hand, (1-c), which is truth-conditionally nearly equivalent to (1-a), is very different in the targeted inference it licenses.

(1) a. #It's almost dark, so turn on only your sidelights.
b. #It's dark, so turn on only your sidelights.
c. It's not yet completely dark, so turn on only your sidelights.

Example (2) shows as well that two seemingly contradictory statements can be part of the same utterance and serve the same conclusion.

(2) A : Is the dinner ready ?
B : Yes, almost.

Once again, if the dinner is *almost* ready, then it is *not* ready, but that does not prevent *B* to present this state of affairs as being identical to the one of being ready, at least for the purpose of that particular context.

Based on examples like (1) and (2) it has been argued that the concept of *argumentation* reveals a dimension of discourse that is distinct from and orthogonal to truth-conditions.

Argumentation *theory*, a development which focuses on and seeks to account for such types of effects, has developed two central hypotheses: first, speakers always speak to a point, such that all conversation is argumentative; and second, that argumentative properties are hardcoded in the grammar of natural languages. Such a theory makes clear that argumentative goals must be somehow present and modeled in a framework of dialogue; the whole question is *how* precisely this should be done.

Argumentation can be explicated by the notion of *Relevance*, for example as in the probabilistic Discourse Theoretic Semantics framework of Merin (1999) (or in comparable ways, e.g. see van Rooij (2004)). Merin framed his work in decision theory, which is a subpart of game theory.

Merin states that a sentence E positively (resp. negatively) argues for a conclusion H if and only if the probability of H after learning E is raised (resp. lowered). The higher the (positive or negative) effect on the probability of H, the more relevant is the uttered proposition E. A proposition is relevant to a *question* if it is relevant to at least an element of π_Q (the partition induced by the question in Hamblin style semantics). A question is relevant to another question if at least one element of the partition induced by the first is relevant to at least one element of the partition induced by the second.

Depending on the discourse situation the argumentative goal might be explicit (e.g. after an imperative), or needs to be induced somehow from the semantic meaning of the speaker's utterance. According to Merin, this is one part of what conversation is all about. Inducing such a goal amounts to finding the proposition H^* in the set of all propositions for which the uttered proposition E is an argument, i.e. $H^* \in \{H|r_H(E) > 0\}$ (where $r_H(E)$ is the relevance of E to H). Decision theoretic considerations on the argumentative goal can be used to derive a variety of pragmatic effects (e.g. conversational and conventional implicatures, presupposition...)

In such a framework, the relevance of an utterance is always defined with respect to an argumentative goal. It is, however, not quite clear whether the argumentative goal is a proposition or a disposition to act (of one of the discourse participants). As far as Merin is concerned, it seems that he identifies the fact of believing a proposition with the disposition to act in a specific way, citing Ramsey (1926) as inspiration.

1.2 Argumentation and Game Theory

At first sight, the question of how to represent argumentative goals may look like somewhat orthogonal to what *is* relevance, and that this is rather about what it means to believe a sentence. However, we do not think that these issues are entirely unrelated, given our initial question of whether argumentation could or should be treated as a part of pragmatics.

The theoretical issue at stake can be stated as follows: if an argumentative goal is to be modeled as a disposition to act of the discourse participants, relevance can be an instance of the expected utility of the speaker in the usual game-theoretical sense, that is, an indication of speaker preferences that will further dictate their actions, assuming rational behaviour. By uttering a sentence, the speaker will induce (under certain circumstances) the addressee to react in a certain way, and the utilities of a given sentence would be the utilities the speaker receives from the (re)action of the addressee upon hearing that sentence.

Conceived in such a way, relevance and expected utility would be two instances of the same phenomenon, and fall into what one may call the *use of language* in a broad sense. In any way, relevance falls out of the *langue* under such a perspective. If the argumentative goal is to be modeled as a proposition, the identification of relevance and expected utility is not possible. But this leaves open the possibility that relevance (and argumentation) are part of the *langue*. As such, it is rather clear that relevance should not be directly encoded in a sentence. What we are looking for are rather "relevance-conditional" semantics, that impose conditions on relevance, than direct reference to pay-offs. This is the same as truth-conditional semantics, which does not deal directly with truth or falsity of propositions, but the conditions which would render them true or false.

Maybe surprisingly, given his stance on belief, Merin (1999) is quite explicit about the technical side of the issue: for him, relevance is a relation between two propositions and an epistemic state (what he calls the 'epistemic context'). The two propositions involved are the argumentative goal H and the proposition that is uttered, E. The epistemic context provides a probability distribution over propositions, including the argumentative goal H.

Merin's way of dealing with relevance has several highly attractive consequences. First, relevance as a relation between propositions given a certain context makes it possible to conceive that items in the *grammar* may be sensitive to such relations, or even manipulate, constrain or order the possible relevance-relations between an uttered proposition and a given or inferred proposition H. Thus, it becomes possible to conceive argumentative properties as based in the *langue*, rather than as being by-products of the *use* of a language. Such a position seems to be necessary in order to account for a certain type of data (e.g., *almost*, as we have seen above, and adversative conjunctions, cf. below). There is thus a good reason for grounding relevance in purely epistemic terms, as far as it enables one to make predictions on the type of utterances. Nevertheless, some properties related to token utterances cannot be accounted for by epistemic means alone, and we argue that those context-specific issues are the ones that benefit from a game-theoretic approach based on a notion of utility distinct from the epistemically based notion of relevance advocated by Merin.

Second, the fact that propositions can be assigned a quantified amount of relevance allowed Merin (2003) to develop relevance-scales, which are able to cover more empirical ground with respect to implicatures than traditional entailment-scales (e.g., like the ones proposed in the neo-Gricean tradition, notably by Horn (1989)).

Having said that, is such an approach conclusively opposed to a direct encoding of expected utilities of the speaker? No. Merin's position is not inconsistent. One may indeed still identify the utility of the speaker with the relevance of a proposition E to a given goal H: believing H (or coming closer to it) will influence the disposition to act of the players, from which one can derive straightforwardly their utilities. However, assuming this identity makes it difficult to imagine how a grammatical item (belonging to some *langue*) should manipulate such a volatile and contextual entity such as a given goal H. As already mentioned, argumentation and relevance do have proper linguistic characteristics, that bear on the felicity and interpretation of utterances. We will examine this proper linguistic characterisation further in the next section with respect to the issue of discourse topics.

While relations between propositions are quite standard elements in a semantician's toolbox, it is not an obvious move to state a grammatical relation between an utterance and a disposition to act. A proposition, whatever may be the precise way in which one wishes to define it, clearly is a linguistic entity. A disposition to act is not a linguistic item, and it is delicate to assume that there could be words sensitive to such an entity [2].

[2] One could certainly model a disposition to act as a set of propositions, but this is a way of not addressing the basic problem.

Setting aside the theoretical issue of what one's position may be with respect to the technical side of representing the argumentative goal H, there is an empirical issue with respect to H, namely that H is not always expressed *expressis verbis*. Yet, figuring out H plays a considerable role in establishing the contextual meaning of a sentence. The question that needs to be addressed is thus the following: are there any rules (or constraints) in the inference of H ? And can we link this process to other processes in discourse interpretation, like the inference of a discourse topic ?

2 The Point of Talking

As is well known, what is said often stands in a rather indirect relation to what is meant. The following example (adapted from Beyssade & Marandin (2002)) illustrates this:

(3) A : Who has been invited ?
B : The post is on strike.
A : Alright, we cancel.

Pragmaticians like Grice (1975) have insisted on the fact that figuring out out what the speaker had in mind is a central element in determining the contextual meaning of a sentence. This procedure of second-guessing the speaker's intention is certainly not grounded in the grammatical rules of a language, and is clearly pragmatic in nature. This 'point of talking' has been called the discourse topic.

Most often, the notion of Discourse Topic (now *D-Topic*) has been worked out from a different perspective than the notion of an argumentative goal.

However, the two concepts have in common that they somehow embody the point of an utterance: either the (possibly implict) question to which an utterance answers in the case of D-Topics or the objective the speaker has in mind when asserting his utterance. We will consider both types of approaches in the remainder of this section. First, we will give a brief reminder about the notions of D-Topic and argumentative goal, and then give our technical argument to link the two. We conclude the section with an application of our proposal to adversative conjunction.

Before moving any further, we should mention however that there are out-spoken opponents to the notion of discourse topics, for instance Asher (2004). He advocates against a unified approach to discourse topics, arguing that the various cases requiring a notion of discourse topic have different, and sometimes incompatible, requirements on what a discourse topic should be. It is not our aim to tip the balance towards a unified or multifarious approach to discourse topics, even though we assume that all utterances do have a discourse topic related to their argumentative orientation.

It is, however, worth noting that Asher acknowledges that there are grammatical devices (such as contrastive accents) which help to disambiguate discourse topics. His central claim is that contrastive accents are not sufficient to clearly identify a single topic (e.g. when accents are omitted). We share the same intuition and our proposition is that other cues are available to single out a discourse topic. However, the crucial point we wish to make is the following: the pragmatic inference of a D-Topic is not only costly, but extremely difficult; grammatical means of restricting the search space will therefore be more than welcome.

2.1 Discourse Topics

Inspired by the works of Roberts (1996), Büring (e.g. Büring (2003)) represents the structure of discourse as a tree whose nodes are *moves* representing declarative or interrogative sentences. This, he calls a *D-Tree*. A grammar of discourse is then a grammar of well-formed *D-Trees*. Crucially, questions in a *D-Tree* can be implicit and accommodated. This allows for a close match between the informational structure of an utterance and the question(s) it answers in a *D-tree*. Chiefly, a declarative sentence must be congruent to its question, otherwise the *D-tree* containing it is ill-formed. The informational structure of an utterance is often represented in the form of a question, and thus, congruence is here defined by comparing the informational structure of the actually uttered utterance with the *D-tree*.

More precisely, *Contrastive Topic* (now: CT) accents and *Informational Focus* accents are linguistic marks that indicate the exact *strategy* used by a speaker. When an utterance is uttered out of the blue, these marks indicate which questions a hearer has to accommodate in order to figure out the speaker's strategy [3]. In the case of overt question, the resulting tree has to match the actual question. The exact marking of these features differs from one language to another: in English *CT*-marking is done with a *B-accent*, whereas in French it is done with a *C-accent* (see Marandin et al. (2002)).

The classical example (4), reproduced in Büring (2003), illustrates the effect of *CT* in discourse.

(4) FRED$_{CT}$ ate the BEANS$_F$
   a. **Step 1:** What did Fred eat ?
   b. **Step 2:** What did Fred eat ? What did Mary eat ? . . . = Who ate what ?

Example (4) is first interpreted as an answer to a question obtained by replacing the informational focus by a wh-word, i.e. (4-a). Next, *CT*-marking indicates a more complex strategy, *i.e.* that the first question is subsumed by a larger one, given by an abstraction of the first question over the *CT*-marked item: (4-b). This will lead us to the representation of a *D-Tree* as follows:

(5)

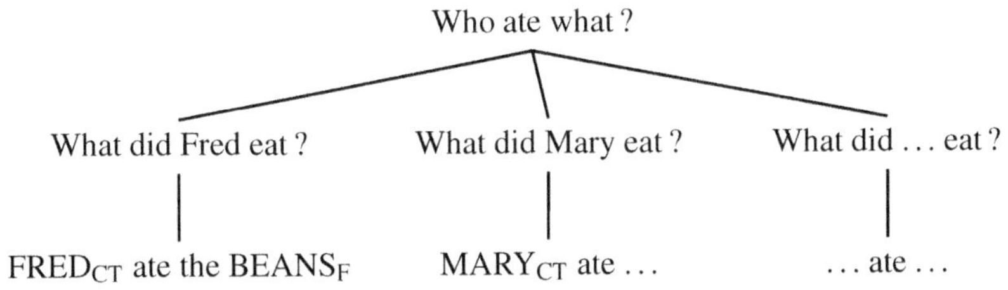

In a nutshell, Büring's approach is one that tries to constrain the possible discourse topics one could infer for an utterance, by using the informational structure of that utterance.

[3] The speaker strategy is the set of questions that subsume the utterance in the *D-tree*.

2.2 Argumentative Goals

The notion of Discourse Topic is also used in argumentation theory, albeit in a different way.

Merin (and works in the argumentative perspective in general) considers a special case of non-cooperative discourse situation, characterized by speaker and addressee having opposite preferences regarding a given goal H. In this case, Merin calls the partition $\{H, \bar{H}\}$ the *issue* of the game and he calls H the *Discourse Topic*. To avoid confusion between theories, we will name the discourse topic in such a setting the *Relevance-Topic (R-Topic)*. The *R-Topic* is meant to be the descriptively most convenient element of the bipartition and is intended to be a neutral term regarding the participants. Given an utterance E of the speaker, the *Protentive Speaker Meaning* (PSM) is that element of the issue for which E is an argument.

Although most argumentative theories deal with non-cooperative discourses, this does not need be so, as acknowledged in Merin (1999). When participants do not have opposite preferences, notions of *PSM* and *Relevance Topic* still have intuitive sense and we extend the definitions to the general case. By *PSM* we will thus mean the goal supported by the speaker, independently of the preferences of the addressee. What we intend to show is that the *R-Topic* is related to the notion of *D-Tree*, and that the two theories have independent contributions to the establishment of the general notion of *Discourse Topic*.

2.3 Linking the Two Perspectives

At this point of our discussion, we will adopt without discussion the view of communication as conceived of in argumentation theory, namely that linguistics items are related by argumentative relations, and that, in speaking, speakers argue for certain conclusions.

In the argumentative perspective, the dichotomic bi-partition $\{H, \bar{H}\}$ is reminiscent of the semantics of polar questions in a Hamblin-style approach[4].

In a non-dichotomic situation, similar parallels can be drawn between a set of possible argumentative goals $\{H_0, \ldots, H_n\}$ and the representation of the meaning of questions as sets of propositions.

Hence, instead of taking the most descriptively convenient element of the partition as the *R-Topic*, we propose to identify the *R-Topic* with the question whose semantics are the set of argumentative goals.

Büring's approach can thus be refined in the following way. When accommodating a question in a *D-Tree* it seems plausible to assume that the question must reflect the argumentative properties of the utterance from which it is derived. More precisely, the inferred question is the question whose semantics match the partition induced by the argumentative situation at hand, that is, the *strategy* (in Büring's sense) must contain the *R-Topic* inferred from the argumentative properties of the sentence.

[4] This is actually independent of the choice of representation for questions. A similar point can be made in a structured meaning approach to questions. The set $\{H, \bar{H}\}$ would be obtained by applying each member of the restriction to the background of the question.

Given that, usually, utterances are not congruent to the *R-Topic* one infers from them, the *R-Topic* needs to be accommodated at the top of the *D-Tree*, thus indicating the starting point of the speaker's strategy.

In the case of overt polar questions such as (6-a), no accommodation would be required.

(6) a. A : Do you like beer ?
b. B : I do

From (6-b), one can infer the *R-topic*: $\{B$ likes beer, B doesn't like beer$\}$ which matches the question in (6-a). Furthermore, (6-b) meets the requirements of congruence imposed by its information structure, which means that the whole dialogue is predicted to be fine without any accommodated strategy.

A more complex example that illustrates our point involves adversative coordination, as marked by the connective *but*. Adversative coordination is often treated in argumentative terms. As observed in Anscombre & Ducrot (1977), sentential *but* connects two propositions E and F such that $r_H(E) > 0, r_H(F) < 0$, i.e. such that they are arguments for opposite conclusions, and thus establishing a dichotomic situation.

(7) This ring is beautiful, but it is expensive.

It is not clear which exact strategy should be inferred from (7) in Büring's framework alone. Probably, what should be inferred is a multiple polar question such as *Is this ring beautiful and is it expensive ?* Yet, this does not account for the presence of *but*, nor for the intuitive *R-topic* that comes out of the argumentative properties of (7).

According to our proposition the *R-topic* is constructed out of the argumentative orientations of each segment. These do not stem from lexical properties but depend on the discourse situation. In our example (7), H', that is, the question which is forming the *R-topic*, must belong to the set of propositions satisfying the argumentative configuration of (7), i.e. $H' \in \{H|\text{sign}(r_H(E)) \neq \text{sign}(r_H(F))\}$. Out of all the propositions in the set, H' is chosen as the one with the highest values of relevance. In the context of (7) a likely candidate for H' would be the purchase of the ring and the *R-topic* would then be *"Should we buy this ring ?"*. The first segment of the utterance would be understood as supporting an affirmative answer to the question, while the second segment supports a negative answer.

Adding the *R-topic*, and allowing in this way for enriching the final *D-tree*, leads to a better representation of the actual meaning of the utterance in (7).

A similar observation has already been noted in Umbach (2005) for example such as (8).

(8) What did John and Mary do ? (Did they do the same ?)
a. $[\text{John}]_{CT}$ $[\text{prepared the dinner}]_F$, but $[\text{Mary}]_{CT}$ $[\text{washed the dishes}]_F$.

Umbach's proposal is that *but*-coordinations with *CT*s answer (negatively) an implicit *quaestio* asking whether both *CT*s are subject to the same predication (indicated in parenthesis in the example). The exact working out of this particular *R-Topic* is feasible, but too complex to fit in this space. What is interesting is that in this case informational structure helps to disambiguate the argumentative goal of the utterance, rather than the other way around (as in (7)).

2.4 Intermediate Summary

So far we have shown how to link argumentative approaches to information structure based accounts of discourse.

Our claim is that the argumentative properties of an utterance constrain the accommodation of discourse strategy, just like the presence of a *CT*-accent does. However, while *CT* define a strategy rather non-ambiguously, argumentative constraints are broader in the sense that they do not identify a single discourse topic. What is interesting about the latter constraints is that they apply to any utterance, and help to retrieve meaning related to speaker's intentions.

3 Where Relevance and Utility Need to be Kept Apart

In the previous section, we have addressed the question of the relation of *R-Topics* and *D-Topics* in the inference of the point of talking. We have argued that *relevance* in Merin's sense is encoded in the grammatical properties of some linguistic items, and allows to pin-point the speaker's strategy with greater accuracy than by recurring to information-structural criteria alone.

We did not use the game-theoretic notion of an expected speaker/hearer utility — or game-theory for that matter —, since relevance on its own could account for everything required in the examples discussed, and since our discussion did not focus on the actual behaviour we would expect from discourse participants. For instance, in some cases, though an addressee is very well able to identify the speaker's strategy, he will want to modify it, and not in the strictly opposing way of partitioning assumed in Merin (1999).

Therefore, in the present section, we will drop the requirement that all instances of conversation are conflictual, that is, assimilable to strictly opposing zero-sum games. Instead, we will suppose with van Rooy (2001) that conversation may take place in a purely cooperative, strictly opposing, or in a mixed setting.

In the remainder of this section, we will study some of these settings. Relevance and expected utility may diverge, and even conflict. More precisely, we will look at situations where the notions of relevance and utility both play a role, albeit a different one, in order to establish a discourse topic or unify the properties of the utterance with an overt discourse topic.

3.1 Bi-partisan Relevance

Abandoning the idea of a strictly opposing setting, it becomes necessary to define a special kind of relevance, that one may call *bi-partisan relevance*. The idea is the following: bi-partisan relevance is the absolute (Merin-)relevance of a proposition with respect to the discourse-topic (identified to H). Assume for instance that speaker and addressee want to know whether H is the case, and that neither of them has any particular stake with respect to H or $\bar{H}$. In such a (purely cooperative) setting, the relevance of a proposition E is the greater, the more it drives to either H or $\bar{H}$. That is, if confronted with the choice between an E_1, which would be positively pertinent to a degree $r_H(E_1) = n$, and an E_2, which would be negatively pertinent to a degree $r_H(E_2) = -k$ with $|k| > |n|$, both speaker and addressee should prefer E_2. The reason is that E_2 allows to sort out the issue with far greater accuracy. Clearly, in such a fully cooperative context, and with two sentences arguing for opposite conclusions, the directionality of the relevance-relation is not helpful, and must be neutralized.

As an example consider (9), and imagine that both A and B are honest law-abiding policemen, whose unique goal is to establish the identity of Sue's killer.

(9) *A :* Did John kill Sue ?

a. He was the last one to see her = E

b. He was in Tokyo at the time of the murder = F

H can be considered to be set by A's utterance to *John killed Sue*. Then, it will be the case that $r_H(E) > 0$ and that $r_H(F) < 0$, i.e. that E and F argue for oppposite conclusions. Furthermore F appears as a better argument for $\bar{H}$ than E for H. So we have: $|r_H(F)| > |r_H(E)|$. In that case, if B knows that E and F are both true he should utter F given that it resolves the issue better than E, even though E is relevant to the issue.

3.2 Changing Goals

The fact of teasing apart relevance and utility furthermore allows us to give an analysis of situations where the inference of an argumentative goal is subject to modification because of the participants' preferences.

Consider the dialogue in (10).

(10) (At the restaurant, after receiving the tab)

a. *A :* I know someone from Austria who will lend me 5 euros.

b. *B (himself Austrian) :* Yes, Kurt will be happy to do it.

Let's assume that the intended goal of A's utterance is H_0 =*B will lend 5 euros to A*. Let's further assume that participants preferences are represented by the amount of money they gain (positive preference) or lend (negative preference) in the described situation. Hence B's preference for H_0 is -5. Among the compatible argumentative goals of (10-a) is H_1 =*Kurt will lend 5 euros to A*. B's preference for H_1 is 0 which means that he prefers it over H_0. Ultimately, A is neutral between H_0 and H_1, and hence he's prone to accept the shift in argumentative goal imposed to him by B.

Therefore, in this situation, the argumentative goal is conjointly established by the participants, given their actual preferences in a game-theoretical kind of approach.

3.3 Mixed Motive Games

A last argument against an identification between relevance and expected utility — which would mean, as we have said before, that belief must be reducible to a disposition to act — comes from the considerations of van Rooy (2001) with respect to mixed motive games. In such contexts, it may occur that, even though somebody believes some information to be true, he will not act according to this belief.

The setting of the games he considers is a quite different one from what we have seen so far, which were signalling games. In this precise case, we are facing a game with communication preceding it, and the result of the communication may influence the pay-offs of speaker and addressee in the game that follows.

Van Rooy (2001) stresses the fact that the common ground (i.e. the set of mutually, and generally accepted information) does not just serve to determine the choice of words of the speaker, and the choice of interpretation of a hearer. The common ground also determines the actions of the discourse-participants in a game that follows the exchange of information.

The scenario goes as follows: a rational speaker will assert a proposition E — which means: risking E to become part of the common ground, and as such influencing the established payoffs for the speaker and addressee — only if E is as least not defavorable to his expected utility. But the addressee has a strategy, too: he will not accept an utterance to become part of the common ground if this reduces his expected utilities. This holds even if he believes the asserted information to be true. And clearly, the addressee must play the resulting game with respect to his old beliefs, because otherwise he would jeopardize the expected utilities.

Let us consider an (invented) example. In the fictive country of Franconia, the prime minister and the minister of the interior compete for the office of the president of Franconia. The prime minister is sent a listing claiming that the minister of the interior has a substantial amount of money in a foreign bank. The prime minister therefore sends the secret service to investigate this claim. Assume that the situation is the following: it would be greatly damaging the campain of the minister of the interior if he had such a bank account, and therefore, a great help for the prime minister. Some months later, the prime minister receives the message:

(11) The minister of the interior does not have a bank account in this foreign bank.

As a rational player, the situation being as put above, the prime minister should not take into account this information, since it would harm his expected outcome in the presidential election.

Now, the proposition in (11) is highly relevant (depending on the person, positively or negatively), in that it might provoke a considerable shift in the outcome of the game. Yet, one must see that it does not affect a preceding disposition to act in any way.

Therefore, such a configuration again shows that belief in a game-theoretic setting cannot be reduced to a disposition to act.

4 Conclusion

In this paper, we have examined the applicability of (parts of) game theory for the semantics (i.e., for the grammatically encoded parts of meaning) of natural languages. We have argued that relevance should be represented explicitly in the grammar of natural languages, and that decision theoretic semantics (as developed in Merin (1999, 2003)) is a possible way to go, making use of a subpart of game theory.

We have also examined the possibility of reducing the phenomenon of argumentation to pragmatics (i.e., purely contextually parts of meaning), and argued against the identification of the concept of *relevance*, as used in argumentative approaches, with that of *expected utility*, as defined in more general game-theoretical settings. Such a reduction could be achieved through the identification of belief with a disposition to act.

Our arguments to oppose such a reductive approach were partly theoretical and partly based on practical considerations regarding the accommodation of a *discourse topic*. We have shown how the two concepts can both contribute independently to the establishment of a given topic, and that from a linguistic perspective, there are good reasons to keep them apart.

Among the perspectives of this work is the integration of our considerations into a formal model of dialogue, especially with respect to relating argumentation and speaker's utility to the concept of *grounding* (e.g. as used in Ginzburg (to appear)) in the case of mutually agreed upon changing goals (as exemplified in section 3.2).

Another direction of research is to find other cues and constraints for the accommodation of discourse topics, that is, what additional elements besides informational structure, argumentation and utility allow us to infer what a speaker wants to convey.

Reference

Anscombre, J.-C., Ducrot, O. : Deux mais en français. Lingua 43, 23–40 (1977)

Anscombre, J.-C., Ducrot, O. : L'argumentation dans la langue. Pierre Mardaga, Bruxelles (1983)

Asher, N. : Discourse Topic. Journal Theoretical Linguistics 30, 163–201 (2004)

Beyssade, C., Marandin, J.-M. : Topic Marking, Discourse Topic and Discourse Moves. In : Workshop on Topics, University of Stuttgart (2002)

Büring, D. : On D-Trees, Beans, and B-Accents. Journal Linguistics and Philosophy 26(5), 511–545 (2003)

Ducrot, O. : Les échelles argumentatives. Les Éditions de Minuit (1980)

Ginzburg, J. : Semantics and Interaction in Dialogue. In : Studies in Computational Linguistics. CSLI Publications, Stanford (to appear)

Grice, H.P. : Logic and Conversation. In : Cole, P., Morgan, J.L. (eds.) Syntax and Semantics. Speech Acts, vol. 3, pp. 41–58. Academic Press, New York (1975)

Hintikka, J., Sandu, G. : Game Theoretical Semantics. In : van Benthem, J., ter Meulen, A. (eds.) Handbook of Logic and Language, pp. 361–410. Elsevier, Amsterdam (1997)

Horn, L.R. : A Natural History of Negation. University of Chicago Press, Chicago (1989)

Marandin, J.-M., Beyssade, C., Delais-Roussarie, E., Rialland, A. : Discourse Marking in French : C Accents and Discourse Moves. In : Speech Prosody 2002, Aix en Provence, France, pp. 471–474 (2002)

Merin, A. : Information, Relevance and Social Decision-Making. In : Moss, L.S., Ginzburg, J., de Rijke, M. (eds.) Logic, Language, and Computation, vol. 2, pp. 179–221. CSLI Publications, Stanford (1999)

Merin, A. : Replacing 'Horn-Scales' by Act-Based Relevance-Orderings to Keep Negation and Numerals Meaningful. In : Forschungsberichte der DFG-Forschergruppe 'Logik in der Philosophie' 110 (2003)

Ramsey, F.P. : Truth and Probability. In : Braithwaite, R.B. (ed.) The Foundations of Mathematics and other Logical Essays, ch. VII, pp. 156–198. Kegan, Paul, Trench, Trubner & Co, London (1926)

Roberts, C. : Information Structure in Discourse : Towards an Integrated Formal Theory of Pragmatics. In : Yoon, J.H., Kathol, A. (eds.) OSU Working Papers in Linguistics. Papers in Semantics, vol. 49, pp. 91–136 (1996)

van Rooij, R. : Cooperative versus argumentative communication. Philosophia Scientia 2, 195–209 (2004)

van Rooy, R. : Relevance of Communicative Acts, Ms., University of Amsterdam (2001)

Umbach, C. : Contrast and Information Structure : A focus-based analysis of but. Journal Linguistics 43(1), 207–232 (2005)

The Geometry and Algebra of Commitment

Felice Cardone

Università di Torino, Dipartimento di Informatica
felice@di.unito.it

Abstract. We propose a formal description, by means of graphical and categorical structures, of mechanisms for handling the dynamics of rights and obligations familiar in jurisprudence. We argue that the formal study of commitment in this setting can contribute new insights to the analysis of a large variety of communicative situations relevant to formal pragmatics.

1 Background and Motivations

This paper is a preliminary description of ongoing research motivated by the desire to express formally the regularities of patterns of interaction that arise in a wealth of communicative situations, e.g., rules of order [18], dialogues in argumentation theory [20] and logic [15], or the exchange of promises that takes place among parties while setting up a contract. We believe that this study is a suitable item in a research agenda for formal pragmatics, primarily as a chapter of the analysis of performatives encompassing not only speech acts but also more general semiotic acts.

The notion of *commitment* has been singled out as a key one for the analysis of interactive situations, from argumentation (see, for example, [20]) to the diagnosis of pathologies of communicative behavior in psychiatry [21]. For example, the notion of *commitment store* in formal dialectics is an accounting device for the obligations (and rights) that the two parties in a dialogue bring into play by performing speech acts like statements and questions: "A speaker who is obliged to maintain consistency needs to keep a store of statements representing his previous commitments, and require of each new statement he makes that it may be added without inconsistency to this store [...] We shall call them commitment-stores: they keep a running tally of a person's commitments" [11, Chapter 8].

We identify commitment with the distribution of rights and obligations across places, and focus on their transformations. Among these, traditional jurisprudence, in particular the Roman law, has singled out the mechanisms of *confusio*, *compensatio* and *delegatio*, which we analyze by reducing them to more general transformations that involve a conservative flow of rights and obligations through a network of generalized systems of accounts expressing a notion of state by means of double-entry recording.

We describe in Section 2 the dynamics of the rights and obligations that bind the participants in an instance of commitment by means of directed graphs

Lecomte and S. Tronçon (Eds.): PRELUDE 2010, LNAI 6505, pp. 147–160, 2011.

representing compositions of basic transformations. Beside a purely arithmetic interpretation of rights and obligations, these graphs also support a more concrete interpretation as describing the flow of documents of a special kind, the *stocks* representing rights to their owners, and the *stubs*, representing obligations.

We introduce in Section 3 an alternative framework where graphs are replaced by morphisms in a category whose objects are systems of accounts. This approach has been stimulated by the algebraic treatment of double-entry accounting given in [7], and connects to recent research on categorical models of computation, especially by Abramsky and his coworkers [1,3,2].

Finally, in the last Section we discuss the relations of the ideas discussed in this paper with the neighboring literature, and outline applications and future developments of our approach.

2 The Geometry of Commitment

We identify the basic form of *commitment* or *obligation* with the existence of a pairing of *places*, the active and the passive end of the obligation. This idea is based on the principle of double-entry accounting, whereby commitments are recorded commitment both in active (right) and in passive (obligation) forms, in different accounts. We represent the existence of such a relation by the presence at each end of one of two types of documents. At the active end, we have the *stock*, at the passive end the *stub*. The terminology comes from the tradition of accounting by means of tally sticks: "the medieval tally was split into two bits of unequal length. The longer (the *stock*, with a stump or handle) was kept as a receipt by the person who handed over goods or money. The shorter [here, the *stub*] was kept by the receiver" [4, page 48]. Other examples of stock include bonds and promissory notes. The use of places is needed as the documents we consider confer rights to their bearer and may lose this property as they change place.

The dynamics of a system of rights and obligations will be identified with the flow of these two types of documents through places, as a result of three kinds of events: (1) creation of $\langle \text{stub}, \text{stock} \rangle$ pairs, (2) destruction of such pairs, and (3) exchange of stock and stub located at different places. Our goal is to find out a way of representing formally the *scripts* that schedule actions of the above three types so as to implement the traditional mechanisms for extinguishing rights and obligations. We do this operationally, by means of graphical structures on which tokens representing stocks and stubs flow.

2.1 Space-Time Diagrams: Definitions

We assume given a set $\mathcal{T}$ of *types*, with a fixed-point free involution associating to each type T a type T^*. Then $\mathcal{T} = \mathcal{T}^+ + \mathcal{T}^-$, where $\mathcal{T}^+ = \{A, B, C, \ldots\}$ is the subset of *positive* types and $\mathcal{T}^- = \{A^*, B^*, C^*, \ldots\}$ is the subset of *negative* types. We also have a set of *places* $\mathcal{P} = \{U, V, W, \ldots\}$.

A *system of accounts* is a string over the set $\mathcal{T} \times \mathcal{P}$, whose elements are pairs TU consisting of a type and a place, where ε denotes the empty string.

The notation $|X|$ denotes the length of the string X. A single *account* (at place U) is then a string over $\mathcal{T} \times \{U\}$. Given a system of accounts X and a place U, we can extract from X an account at place U by means of a projection operation denoted by $X \upharpoonright U$ and defined inductively as follows:

$$\begin{aligned} \varepsilon \upharpoonright U &= \varepsilon \\ (TV)X' \upharpoonright U &= X' \upharpoonright U && \text{if } V \neq U \\ &= (TU)(X' \upharpoonright U) && \text{if } V = U \end{aligned}$$

An account $X = X \upharpoonright U$ encodes a traditional T-account (see [7]) whose *debit* and *credit* part are respectively the left and right columns

$$\begin{array}{c|c} \multicolumn{2}{c}{U} \\ \hline X^- & X^+ \end{array}$$

Given a system of accounts X, the notations X^+ and X^- for the *positive* and *negative* parts of X are defined inductively on X:

$$\begin{aligned} \varepsilon^- &= \varepsilon \\ ((TV)Y)^- &= (AV)Y^- && \text{if } T = A^* \\ &= Y^- && \text{if } T = A \\ \varepsilon^+ &= \varepsilon \\ ((TV)Y)^+ &= (AV)Y^+ && \text{if } T = A \\ &= Y^+ && \text{if } T = A^* \end{aligned}$$

Observe that both X^+ and X^- are strings over $\mathcal{T}^+ \times \mathcal{P}$. The involution defined on $\mathcal{T}$ can be extended to systems of accounts following the same pattern:

$$\begin{aligned} \varepsilon^* &= \varepsilon \\ ((TV)Y)^* &= (T^*V)Y^* \end{aligned}$$

A system of accounts X, as a string, has an associated set $\mathtt{Pos}(X)$ of *positions*, which is the subset of $\mathbb{N}$ defined inductively by the clauses:

$$\begin{aligned} \mathtt{Pos}(\varepsilon) &= \emptyset \\ \mathtt{Pos}((TV)Y) &= \{1\} \cup \{n+1 \mid n \in \mathtt{Pos}(Y)\}. \end{aligned}$$

In the opposite direction, for every $n \in \mathtt{Pos}(X)$, we define $X(n)$ as the element occurring at position n in X:

$$\begin{aligned} ((TV)Y)(1) &= TV \\ ((TV)Y)(n+1) &= Y(n) && \text{if } n \in \mathtt{Pos}(Y) \\ &\quad \text{undefined} && \text{otherwise} \end{aligned}$$

Pairs from the set $\mathcal{T} \times \mathcal{P}$ are used to label the edges of oriented graphs whose nodes have one of the forms shown in Figure 1, where U and V are places:

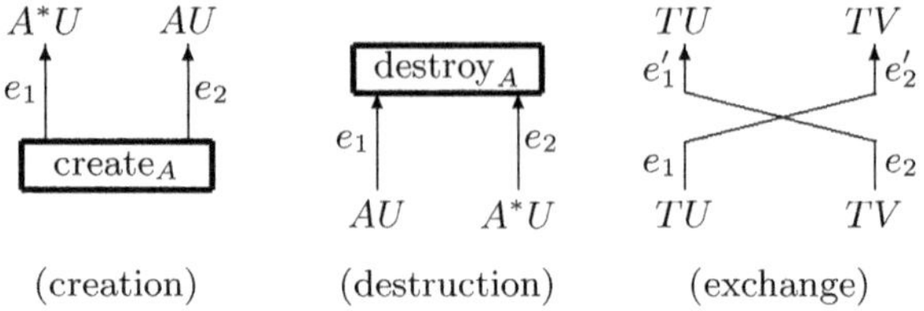

Fig. 1. Nodes, edges and their labels

Intuitively, create_A represents the creation of a pair $\langle \mathrm{stub}, \mathrm{stock} \rangle$, with the stub (of type A^*) and the stock (of type A) located at the same place U. Dually, $\mathrm{destroy}_A$ represents the annihilation of a stock of type A and a stub of type A^* meeting at the same place U. An exchange node represents the trading of a stock of type T at place U for a stock of the same type at a different place V.

Larger graphs, that we shall call *space-time diagrams*, are built from primitive nodes inductively, by connecting each outgoing edge of one space-time diagram to at most one incoming edge (labeled by the same type and place) of another diagram, and symmetrically for incoming edges. (In the sequel, when drawing space-time diagrams, we shall freely cross wires located *at the same place* whenever this is necessary in order to match types: such a crossing should not be confused with an exchange node, which instead involves wires at different places.)

2.2 Executing Space-Time Diagrams

Space-time diagrams are meant to describe the motion of stocks and stubs through places whose states are (recorded as) systems of accounts. For example, the space-time diagrams depicted in Figure 2(b),(c) describe two different transformations whose input state is a system of accounts

(stocks on the right columns). Both transformations do not alter the state; however, diagram (b) corresponds to the identical transformation, whereas diagram (c) exchanges the stocks on the incoming edges.

We can execute space-time diagrams in a way that resembles the token game played on Petri nets. Indeed, by regarding the nodes as transitions and edges as places, every such diagram becomes a special kind of condition/event net, see [17], where every condition has at most one incoming and at most one outgoing event: the resulting net can be represented as an oriented graph whose arcs are the conditions and whose nodes are the events.

A *marking* of a space-time diagram Δ is an injective assignment m of at most one *token* to the edges of Δ. In particular, a diagram Δ with n incoming edges $i_1, \ldots, i_n$ and m outgoing edges $o_1, \ldots, o_m$ has an *input marking* m_{in} assigning to edge i_k token τ_k. The notion of enabling of a node in a marking, the firing

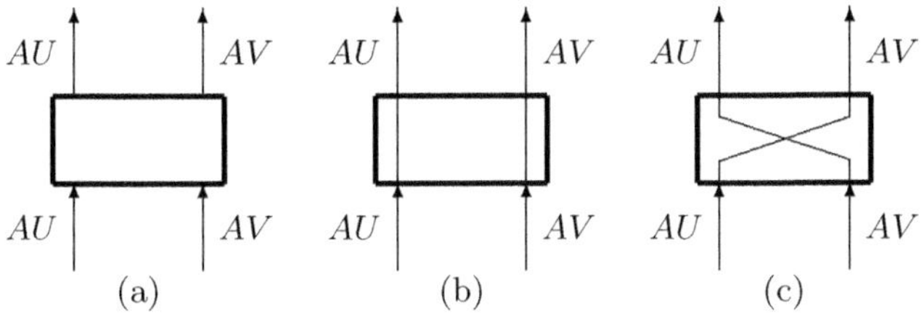

Fig. 2. Two space-time diagrams describing different scripts of the same type

rule and the notion of reachability of a marking from another marking are the same as for condition/event nets [17, §2.1].

We want now to make precise the straightforward notion of identity of a token through a sequence of firings. To this end, let m be a marking of Δ, and $\tau = m(e)$ for some edge e. For every sequence of firings σ starting at m, we define the *residual of* τ *after* σ (written τ/σ), when it exists, by induction on the length of σ:

- if e is not the input edge of any node, then $\tau/\sigma = \tau$;
- otherwise, let e be the input edge of a node ν.
  - If ν is never fired in σ, then $\tau/\sigma = \tau$;
  - otherwise, there is a step $\overline{m} \to \underline{m}$ in σ in which node ν is fired. We consider the possible forms of ν, according to Figure 1:
    * (exchange) assume that $e = e_1$, the other case is symmetric: $\overline{m}(e_1) = \tau$ and $\underline{m}(e_2') = \tau$, therefore τ/σ is again the residual of τ after the final segment of σ that starts at $\underline{m}$ (if this exists);
    * (destroy$_A$) there are three cases:
      (a) ν occurs in the context shown in Figure 3(i). Observe now that, in order for ν to be enabled, the node create$_A$ must have fired earlier giving origin to a marking m^* in σ from which $\overline{m}$ is reachable and such that $m^*(e'') = \tau'$ for some new token τ'. Then τ/σ is the residual of τ' after the final segment of σ that starts at m^* (if this exists).
      (b) ν occurs in the context shown in Figure 3(ii). This case is the symmetric of the former.
      (c) Otherwise, τ has no residual after σ.

The preservation of the identity of tokens through the transformations described in Figure 3 corresponds to the disciplined copying of a document by destroying the original. The residual of a token τ of the input marking m_{in} of Δ after firing all nodes in Δ (in any order) is written simply τ/Δ. Observe that for every space-time diagram Δ, every token has at most one residual after σ, and residuals after σ of distinct tokens are distinct, for any firing sequence σ.

Definition 1. *Space-time diagrams Δ, Δ' with incoming edges $i_1 \ldots i_n$ and outgoing edges $o_1, \ldots, o_m$, labeled by pairs $T_1U_1, \ldots, T_nU_n$ and $T_1'V_1, \ldots, T_m'V_m$,*

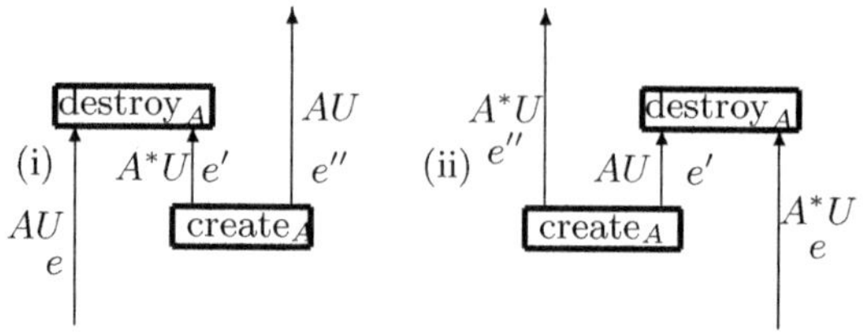

Fig. 3. Contexts for regenerating a token

respectively, are equal *if, for every incoming edge i_k, $m_{in}(i_k)/\Delta$ exists and is on edge o_ℓ if and only if $m_{in}(i_k)/\Delta'$ exists and is on edge o_ℓ.*

Equality of space-time diagrams is almost trivial, due to the very special nature of the nodes. It holds whenever one input token has residuals on corresponding outcoming edges of the two diagrams. With this definition, for each place U, the diagram in Figure 3(i) is equal to a single edge labeled AU, whereas the diagram in Figure 3(ii) is equal to a single edge labeled A^*U. However, the two space-time diagrams shown in Figure 2(b),(c) turn out to be different: any input token has residuals on different outgoing edges after executing each diagram.

2.3 *Delegatio*, *Compensatio* and *Confusio*

We shall discuss the intuitive interpretation of space-time diagrams and show their uses in modeling the interactions prescribed by the traditional mechanisms for the transfer and extinction of obligations. We shall interpret stocks and stubs of type A as tokens of type A and A^*, respectively. A type is assumed to characterize completely the content of an obligation. Let us consider the following example from [4, page 54], where all obligations involved are of type A:

> If V owes Z, but is owed by U, let him – V – make out a receipt to U and give it to Z, and let Z not part with it till he receives the money.

We look at these actions as a dialogue involving three places, U, V and Z, in which the speech acts are replaced by exchanges of $\langle$stub, stock$\rangle$ pairs. Observe that this mechanism coincides with delegation in Roman law (according to Book 46th of the Digest of Justinian, “Delegare est vice sua alium reum dare creditori vel cui iusserit”). The initial state of this series of transformations is represented by the system of accounts

U		V		Z	
A		A	A		A

,

the final state is

U		V		Z	
A					A

.

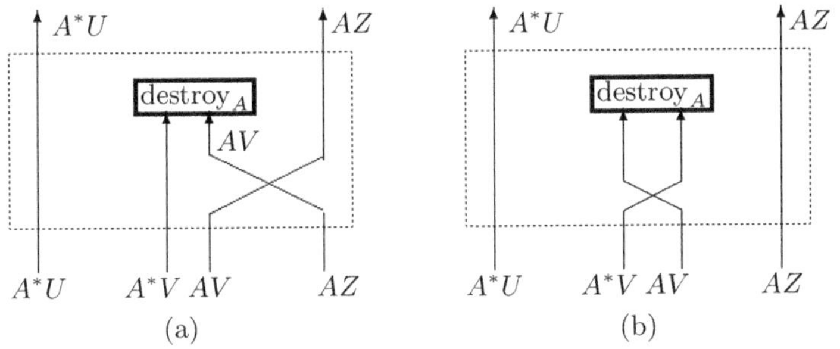

Fig. 4. Scripts for delegation

Delegation is implemented by the space-time diagram in Figure 4(a) or, alternatively, by that in Figure 4(b). Observe, however, that only the first diagram describes the flow of stocks and stubs as in the example.

Other mechanisms from the Roman law of a similar nature include *confusio* and *compensatio*. The former refers to the situation where the position of debtor and creditor with regard to one and the same obligation merge in one person [22]. In our view, this is expressed by the meeting at one place of a stub and a stock of the same type, which can then be annihilated, as in the relative script which coincides with one application of destruction.

Similarly, compensation can be applied to a situation where two obligations of reverse sign are simultaneously present, represented by the system of accounts

$$\frac{U}{A \mid A} \qquad \frac{V}{A \mid A}$$

leading to the situation in which U and V are empty. This transformation can be implemented in the two ways illustrated in Figure 5.

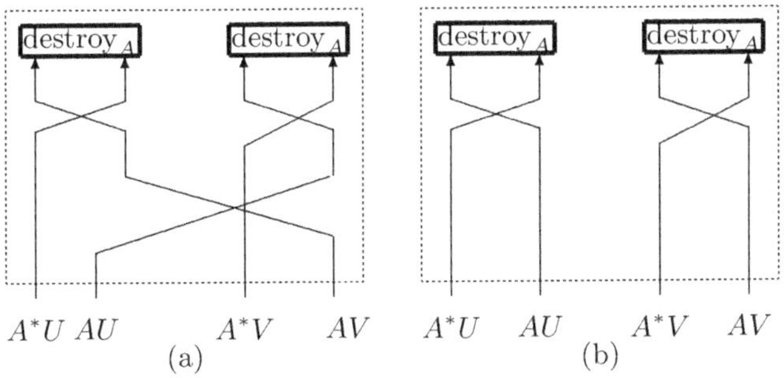

Fig. 5. Scripts for compensation

3 The Algebra of Commitment

We study now the algebraic structure of systems of accounts. In order to define a category of systems of accounts, we exploit the familiar construction of a compact closed category from a traced monoidal category [12], in a variant devised by Abramsky [1] as a model of both Girard's Geometry of Interaction [10] and of a simple form of game semantics. However, we first need to describe further operations on systems of accounts.

Definition 2. *1. The* tensor product $X \otimes Y$ *of systems of accounts* X *and* Y *is their concatenation,*
2. the dual *of a system of accounts* X *is the system of accounts* X^**,*
3. the system of accounts 0 *is* ε*.*

Intuitively, the operation X^* exchanges credit and debit parts of all accounts of the system X, i.e., $(X^*)^+ = X^-$ and $(X^*)^- = X^+$. The operations on systems of accounts just introduced satisfy an elegant equational theory: for example, $(X \otimes Y)^* = (X^* \otimes Y^*)$. By the same argument, we also have the equations $(X \otimes Y)^+ = (X^+ \otimes Y^+)$ and $(X \otimes Y)^- = (X^- \otimes Y^-)$. In the present context we are, however, more interested in a congruence relations that captures the equality of balance on accounts, whose formal description is our next topic.

Definition 3 (Matching)

1. *Given strings* X, Y *over* $\mathcal{T}^+ \times \mathcal{P}$, *a* matching f of X with Y *is a bijection* $f : \mathtt{Pos}(X) \longrightarrow \mathtt{Pos}(Y)$ *such that, for all* $p \in \mathtt{Pos}(X)$, $X(p) = TV$ *if and only if* $Y(f(p)) = TU$.
2. *Two matchings* f, f' *of* X *with* Y *are* equal *if and only if, for every* $p \in \mathtt{Pos}(X)$, $Y(f(p)) = Y(f'(p))$.

A matching of X with Y is then just a way of associating positions of Y to positions of X, bijectively and in such a way as to preserve the types (but not necessarily the places).

3.1 Matchings as Morphisms

We consider a category $\mathbb{M}$ whose objects are strings over $\mathcal{T}^+ \times \mathcal{P}$ and whose morphisms with domain X and codomain Y are the matchings of X with Y. This is a monoidal category with tensor product given by concatenation, it is symmetric and, more importantly from the present standpoint, it is traced. Indeed, let $f : X \otimes Z \longrightarrow Y \otimes Z$ be a morphism. Then f is a bijection $\mathtt{Pos}(X) + \mathtt{Pos}(Z) \longrightarrow \mathtt{Pos}(Y) + \mathtt{Pos}(Z)$, and the algorithm below describes exactly the iterative procedure exploited by (a variant of) the Garsia-Milne involution principle [8] in order to construct a bijection $\mathtt{Pos}(X) \longrightarrow \mathtt{Pos}(Y)$:

for any position $p \in \mathtt{Pos}(X)$:
  $q := f(p)$;
  while (q is a position of Z)
    $q := f(|X| + q - |Y|)$;
  return q

Observe that the loop terminates within $|Z|$ iterations, yielding a matching of X with Y. This is the same as the trace of f obtained by regarding f as a partial injective function $\mathtt{Pos}(X)+\mathtt{Pos}(Z) \longrightarrow \mathtt{Pos}(Y)+\mathtt{Pos}(Z)$, [3, §5.1]. Summarizing these observations, we can state:

Proposition 1. *$\mathbb{M}$ is a traced symmetric monoidal category.*

We now recall briefly how a new category $\mathcal{G}(\mathbb{M})$ is built out of $\mathbb{M}$, following [1,3]. The objects of $\mathcal{G}(\mathbb{M})$ are pairs $\langle X^+, X^- \rangle$ of objects of $\mathbb{M}$, a morphism $f : \langle X^+, X^- \rangle \longrightarrow \langle Y^+, Y^- \rangle$ in $\mathcal{G}(\mathbb{M})$ is a morphism $f : X^+ \otimes Y^- \longrightarrow X^- \otimes Y^+$ in $\mathbb{M}$. The identity morphism $X^+ \otimes X^- \longrightarrow X^- \otimes X^+$ is the obvious matching that exchanges positions. Composition in $\mathcal{G}(\mathbb{M})$ is given by *symmetric feedback*, exploiting the traced structure of $\mathbb{M}$, which can be expressed graphically by the diagram:

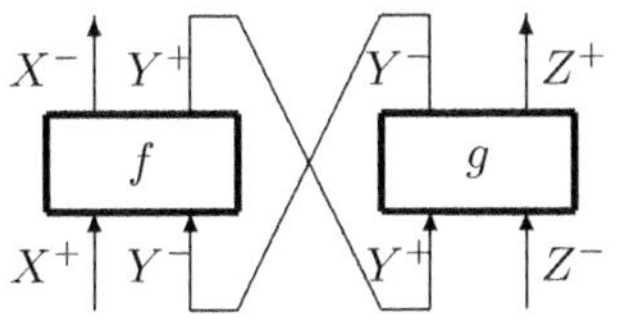

Fig. 6. Composition in $\mathcal{G}(\mathbb{M})$ as symmetric feedback

3.2 A Category of Accounts

A 0-*account* is an account X for which there is a matching of X^+ with X^-. A 0-*system* is a system X of accounts such that $X \restriction U$ is a 0-account, for every place U. The tensor product of 0-systems is again a 0-system, because 0-accounts are closed under tensor product.

Definition 4. *Given systems of accounts X, Y as above, define $X \sim Y$ if and only if $X \otimes Y^*$ is a 0-system.*

The proof of the following properties is mostly straightforward. The proof that $\sim$ is transitive exploits composition of matchings $X^+ \otimes Y^- \longrightarrow X^- \otimes Y^+$ and $Y^+ \otimes Z^- \longrightarrow Y^- \otimes Z^+$ by means of symmetric feedback, which amounts to their composition in $\mathcal{G}(\mathbb{M})$. In this case, X, Y and Z can be taken to be accounts at the same place.

Proposition 2. *1. $X \sim Y$ if and only if $X \otimes Y^* \sim 0$,*
2. $\sim$ is a congruence (w.r.t. the $\otimes$-* structure on systems of accounts),
3. $X \otimes X^* \sim 0$,
4. systems of accounts modulo $\sim$ form an abelian group.

We now define a category $\mathbb{A}cc$ whose objects are going to be systems of accounts, inspired the construction of the Pacioli group in [7], which is an instance of a classical construction of the group of differences of a commutative monoid.

Intuitively, we want the morphisms of our category of systems of accounts to be conservative, i.e., they preserve the balance of every single account in the system. In particular, our morphisms must connect congruent objects.

Definition 5. *Let* $\mathbb{A}cc$ *be the category whose objects are the systems of accounts where, if* $X \otimes Y^* \sim 0$*, a morphism* $f : X \longrightarrow Y$ *is a matching of* $(X \otimes Y^*)^+$ *with* $(X \otimes Y^*)^-$*.*

So, a morphism $f : X \longrightarrow Y$ in $\mathbb{A}cc$ is a morphism $X^+ \otimes Y^- \longrightarrow X^- \otimes Y^+$ in $\mathbb{M}$, and therefore also a morphism $\langle X^+, X^- \rangle \longrightarrow \langle Y^+, Y^- \rangle$ in $\mathcal{G}(\mathbb{M})$.

Assume now that $f : X \longrightarrow Y$ and $g : Y \longrightarrow Z$. Observe first that $X \otimes Z^* \sim 0$ in $\mathbb{A}cc$, by Proposition 2. We define $g \circ f : X \longrightarrow Z$ in $\mathbb{A}cc$ to be the same as the composition of f and g in $\mathcal{G}(\mathbb{M})$. More explicitly, by looking at the diagram of Figure 6 as a flow diagram for computing composition in $\mathcal{G}(\mathbb{M})$, for any position $p \in \mathtt{Pos}((X \otimes Z^*)^+)$, $(g \circ f)(p) \in \mathtt{Pos}((X \otimes Z^*)^-)$ is the position q returned by the algorithm described as follows in the case when $p \in \mathtt{Pos}(X^+)$ (the case when $p \in \mathtt{Pos}(Z^-)$ is symmetrical):

$q := f(p)$;
if (q is is a position of X^-) **then**
  return q;
while (q is a position of Y^+) **do**
  $q := g(q - |X^-|)$;
  if (q is a position of Y^-) **then**
    $q := f(|X^+| + q)$;
  else return q

We can visualize morphisms of $\mathbb{A}cc$ by representing negative types with hollow circles ($\circ$) and positive types with solid circles $\bullet$), omitting their names and the places. Matching types are then represented by links of the form $\bullet\!\!-\!\!\circ$. With this notation we can illustrate schematically the composition $g \circ f$ in Figure 7 in the case when p is a position of X^+ (the lower row represents X, the middle rows Y and Y^* respectively, and the upper row Z^*).

Proposition 3. *Composition of morphisms in* $\mathbb{A}cc$ *is well-defined and is associative.*

We shall not dwell upon the structure of the category $\mathbb{A}cc$ any further, leaving this to future work. With hindsight, it is not surprising that symmetric traced

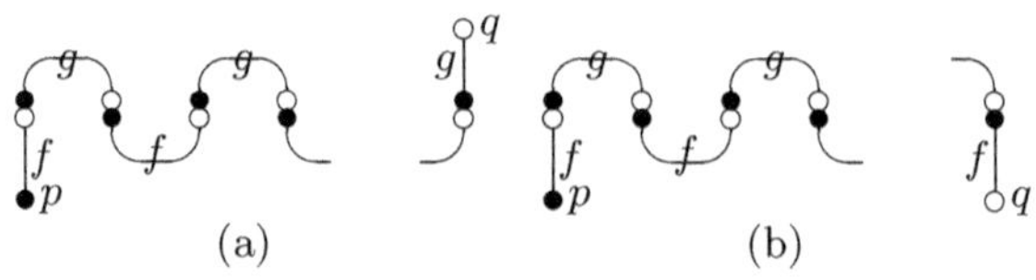

Fig. 7. Two examples for composition

monoidal categories and related compact closed categories like $\mathcal{G}(\mathbb{M})$ enter in the present account. The elementary formalization of double-entry accounting given by Ellerman [7] relates it to the construction of the group of differences from a commutative (additive) monoid generalizing the well-known construction of the integers from the natural numbers. If the monoid is regarded as a discrete symmetric monoidal category, the tensor product being the monoid operation, and if in addition the monoid has the cancellation property, this category is traced monoidal and the construction of the group of differences turns out to be a special case of the Int construction studied in [12], which inspired Abramsky's categorical reconstruction of Girard's geometry of interaction for linear logic [1,10], taking the form of $\mathcal{G}(\mathbb{C})$ for a traced symmetric monoidal $\mathbb{C}$. The relevance of (compact closed bi)categories in modeling double-entry accounting was first pointed out in [13].

3.3 Examples

The following examples describe the interpretation of the space-time diagrams of Section 2.1 as morphisms of $\mathbb{A}cc$. A space-time diagram with incoming edges $i_1 \ldots i_n$ and outgoing edges $o_1, \ldots, o_m$ labeled by pairs $T_1U_1, \ldots, T_nU_n$ and $T'_1V_1, \ldots, T'_mV_m$, respectively, is interpreted as a morphism of $\mathbb{A}cc$ of the form

$$[\![\Delta]\!] : T_1U_1 \otimes \cdots \otimes T_nU_n \longrightarrow T'_1V_1 \otimes \cdots \otimes T'_mV_m.$$

Creation and destruction Observe that $A^*U \otimes AU = (A^*U)(AU)$ is an object for any place U and type A, representing the account

$$\frac{U}{A \mid A}.$$

The only possible matching describes both the morphism $\mathrm{create}_A : 0 \longrightarrow A^*U \otimes AU$, and the morphism $\mathrm{destroy}_A : AU \otimes A^*U \longrightarrow 0$

Exchange Consider the types $AU \otimes AV$ and $(AV \otimes AU)^* = A^*V \otimes A^*U$. Observe that $(AU \otimes AV) \otimes (A^*V \otimes A^*U) \sim 0$, and the matching that associates the occurrence of A at U with the occurrence of A^* at V, and the occurrence of A at V with the occurrence of A^* at U is a morphism, corresponding to exchange.

This interpretation can be extended inductively to all space-time diagrams. As a further example of this interpretation, Figure 8 shows the construction of the morphisms of $\mathbb{A}cc$ associated with the diagrams shown in Figure 3 of Section 2.1. Both are equal to the identities at the respective types (the *zig-zag identities*).

Finally, we state without proof the relation between the operational equivalence of space-time diagrams of Definition 1 and the equality of their interpretations as morphisms of $\mathbb{A}cc$, observing that matchings are just another way of presenting the paths followed by tokens through space-time diagrams.

Proposition 4. *If space-time diagrams Δ, Δ' with incoming and outgoing edges labeled by pairs $T_1U_1, \ldots, T_nU_n$ and $T'_1V_1, \ldots, T'_mV_m$ are equal, then $[\![\Delta]\!] = [\![\Delta']\!] : T_1U_1 \otimes \cdots \otimes T_nU_n \longrightarrow T'_1V_1 \otimes \cdots \otimes T'_mV_m$.*

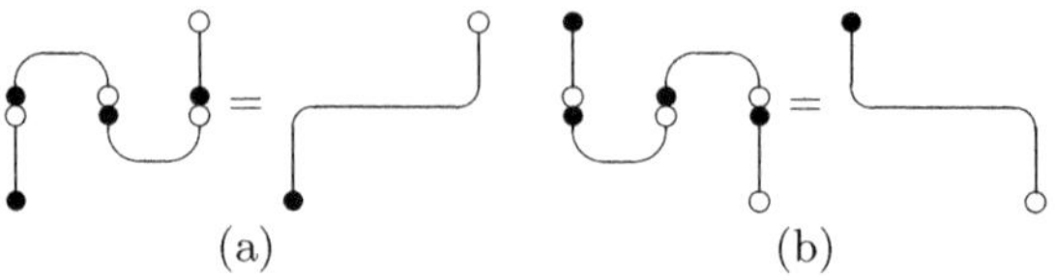

Fig. 8. The zig-zag identities

4 Related Research and Future Work

In order to place our proposal in context, we mention some of the ideas that have been influential on our development. It has been observed in [5,9] that the use of stock and stub as an accounting device, dating back to the prehistory of writing, can be described by means of Feynman graphs, looking at stocks and stubs as particles and antiparticles that can move forward and, interestingly, backwards, allowing a nice description of the possible ways of transferring assets (and liabilities). By means of space-time diagrams we side-step the metric complications entailed by this mechanical model of double-entry accounting.

The idea of having tokens of two kinds on a graph to be executed like a Petri net is similar to what happens in the financial game of [16]. Again, this gives rise in a straightforward way to compact closed categories, as stressed recently in [2]. There, however, the placement of tokens along the graphs does not follow the double-entry principle, which instead lies at the basis of our explicit representation of the interactions that bring about the transfers of rights and obligations.

What we regard as an advantage of our proposal is the simultaneous representation of the states of a system of rights and obligations, which is what is represented traditionally by accounts, *and* the transformations of these states by means of the flow of documents of a simple kind. Furthermore, these two dimensions are linked by an equational theory that includes familiar laws. One area for further work is the complete analysis of this theory and its relations to the graphical calculi for monoidal categories studied in [19] and in the recent categorical approaches to quantum physics [6]. Also the striking similarities between matchings in $\mathbb{A}cc$ and Kelly-Mac Lane graphs [14] need more investigation.

One of the original motivations for the work outlined in this paper was the search for a rationale behind the procedural rules of Lorenzen dialogues [15]. While attack/defense rules for the logical constants have a natural justification, procedural rules have an *ad hoc* character that hinders the foundational interpretation of the dialogical setting. We believe that the arithmetic of commitments that we propose can be extended to the management of the flow of rights and obbligations in a Lorenzen dialogue. While the types used in this paper represent very special kinds of *contracts*, namely unilateral contracts that are either pure rights or pure obligations, we are currently exploring the possibility to enrich this repertoire with contracts of other kinds. In particular, every implication $\varphi \to \psi$ acts, in the dialogical context, as a contract that commits the asserting party to ψ in exchange for the adversary's commitment to φ (with adversary triggering the application of the contract).

As a first, almost trivial but suggestive example, consider the dialogue for the formula $A \to A$ asserted by the Proponent (at place P). When the Opponent (at place O) asserts A he triggers the contract associated to $A \to A$, reaching a stage represented by the system

$$A^*O \otimes AO \otimes A^*P \otimes AP.$$

This is a 0-system corresponding to the initial state of the script for compensation, whose execution yields 0. The justification of the dialogical validity of this formula by means of a procedural rule is replaced in this case by the requirement that the system of accounts arising from this dialogue be a 0-system, where all proof obligations can be extinguished by scripts representing morphisms in $\mathbb{A}cc$. We are currently striving to define a formal setting for associating systems of accounts to dialogues so that validity correspond to the fact that the relevant accounts balance.

Acknowledgements. I would like to thank the referees for their stimulating and encouraging comments. Marco Gaboardi, Michele Pagani, Mauro Piccolo and Luca Vercelli have listened patiently to a long presentation of the ideas sketched in this paper, providing valuable suggestions. Jean-Yves Girard pointed out problems arising from the sloppy use of the notion of occurrence during the presentation of an earlier version of this material in Paris. This research has been partly supported by MIUR Project PRIN07RNC 7.7.2.60 "CONCERTO".

References

1. Abramsky, S.: Retracing some paths in process algebra. In: Sassone, V., Montanari, U. (eds.) CONCUR 1996. LNCS, vol. 1119, pp. 1–17. Springer, Heidelberg (1996)
2. Abramsky, S.: Petri nets, discrete physics, and distributed quantum computation. In: Degano, P., De Nicola, R., Bevilacqua, V. (eds.) Concurrency, Graphs and Models. LNCS, vol. 5065, pp. 527–543. Springer, Heidelberg (2008)
3. Abramsky, S., Haghverdi, E., Scott, P.: Geometry of interaction and linear combinatory algebras. Mathematical Structures in Computer Science 12(5), 625–665 (2002)
4. Baxter, W.: Early accounting: The tally and the checkerboard. The Accounting Historians Journal 16(2), 43–83 (1989)
5. Braun, D.: Assets and liabilities are the momentum of particles and antiparticles displayed in Feynman-graphs. Physica A: Statistical Mechanics and its Applications 290(3-4), 491–500 (2001)
6. Coecke, B.: Kindergarten quantum mechanics (2005), `http://arxiv.org/abs/quant-ph/0510032`
7. Ellerman, D.P.: The mathematics of double entry bookkeeping. Mathematics Magazine 58(4), 226–233 (1985)
8. Feldman, D., Propp, J.: Producing new bijections from old. Advances in Mathematics 113, 1–44 (1995)
9. Fischer, R., Braun, D.: Nontrivial bookkeeping: a mechanical perspective. Physica A: Statistical Mechanics and its Applications 324(1-2), 266–271 (2003)

10. Girard, J.Y.: Geometry of interaction I: Interpretation of system F. In: Bonotto, C., Ferro, R., Valentini, S., Zanardo, A. (eds.) Logic Colloquium 1988, pp. 221–260. North-Holland, Amsterdam (1989)
11. Hamblin, C.: Fallacies. Methuen & Co., London (1970)
12. Joyal, A., Street, R., Verity, D.: Traced monoidal categories. Mathematical Proceedings of the Cambridge Philosophical Society 119, 447–468 (1996)
13. Katis, P., Sabadini, N., Walters, R.: On partita doppia (1998) (unpublished)
14. Kelly, G.M., Mac Lane, S.: Coherence in closed categories. Journal of Pure and Applied Algebra 1, 97–140 (1971)
15. Lorenzen, P.: Ein dialogisches Konstructivitätskriterium. In: Infinitistic Methods, pp. 193–200. Pergamon Press and PWN, Oxford and Warsaw (1961)
16. Martí-Oliet, N., Meseguer, J.: From Petri nets to linear logic through categories: A survey. International Journal of Foundations of Computer Science 2(4), 297–399 (1991)
17. Reisig, W.: Petri nets: an introduction. Springer, Heidelberg (1985)
18. Robert, H.: Roberts Rules of Order. In: The Standard Guide to Parliamentary Procedure. Bantham Books, New York (1986)
19. Selinger, P.: A survey of graphical languages for monoidal categories (December 2008) (unpublished)
20. Walton, D.N., Krabbe, E.C.: Commitment in Dialogue: Basic Concepts of Interpersonal Reasoning. State University of New York Press, Albany (1995)
21. Watzlawick, P., Beavin, J.H., Jackson, D.D.: Pragmatics of Human Communication. In: A Study of Interactional Patterns, Pathologies, and Paradoxes. W.W. Norton & Co., New York (1967)
22. Zimmermann, R.: The Law of Obligations: Roman Foundations of the Civilian Tradition. Oxford University Press, Oxford (1996)

Compliance

Jeroen Groenendijk and Floris Roelofsen

ILLC/Department of Philosophy
Universiteit van Amsterdam
http://www.illc.uva.nl/inquisitive-semantics

1 Introduction

The aim of this paper is to motivate and specify the logical notion of *compliance*, which judges whether or not a certain sentence makes a significant contribution towards resolving a given issue in a cooperative dialogue that is geared towards the exchange of information. We assume that such a contribution may consist in (partially) resolving the issue, or in raising an easier to answer sub-issue (cf. Roberts, 1996). Thus, among other things, compliance will provide a characterization of *answerhood* and *subquestionhood*: it will tell us which sentences count as (partial) answers to a given question, or as subquestions of that question.

Compliance will be defined within the framework of *inquisitive semantics* (Groenendijk, 2008; Mascarenhas, 2008), which departs from partition theories of questions (Groenendijk and Stokhof, 1984; Groenendijk, 1999) in a crucial way: questions are no longer analyzed as partitions of logical space, but as sets of alternative (though possibly overlapping!) 'possibilities'. We will see that, as a consequence of this departure, answerhood and subquestionhood can no longer be defined in terms of entailment.

The paper is organized as follows. Section 2 reviews the basic notions of inquisitive semantics, and points out why answerhood and subquestionhood can not be defined in terms of entailment in this framework. Section 3 introduces the notion of compliance, and section 4 discusses the notion of homogeneity, which captures certain quantitative preferences among compliant responses.

2 Inquisitive Semantics

Classically, the meaning of a sentence is identified with its informative content. Stalnaker (1978) gave this informative notion a dynamic and conversational twist by taking the meaning of a sentence to be its potential to change the common ground, where the common ground is viewed as a body of shared information as it has been established in a conversation.

The notion of meaning that resulted from this 'dynamic turn' reflects the *active* use of language in *changing* information. However, what it does not yet capture is the *interactive* use of language in *exchanging* information. This requires yet another turn, an 'inquisitive turn', leading to a notion of meaning that directly reflects the nature of informative dialogue as a cooperative process of raising and resolving issues.

Lecomte and S. Tronçon (Eds.): PRELUDE 2010, LNAI 6505, pp. 161–173, 2011.

2.1 Propositions as Proposals

We follow the standard practice of referring to the meaning of a sentence as the proposition that it expresses. The classical logical-semantical picture of a proposition is a set of possible worlds, those worlds that are compatible with the information that the sentence provides. The common ground is also standardly pictured as a set of worlds, those worlds that are compatible with the conversational participants' common beliefs and assumptions. The communicative effect of a sentence, then, is to enhance the common ground by excluding certain worlds, namely those worlds in the common ground that are not included in the proposition expressed by the sentence.

Of course, this picture is limited in several ways. First, it only applies to sentences which are used exclusively to provide information. Even in a typical informative dialogue, utterances may serve different purposes as well. Second, the given picture does not take into account that enhancing the common ground is a cooperative process. One speech participant cannot simply change the common ground all by herself. All she can do is *propose* a certain change. Other speech participants may react to such a proposal in several ways. These reactions play a crucial role in the dynamics of conversation.

In order to overcome these limitations, inquisitive semantics starts with an altogether different picture. It views propositions as proposals to enhance the common ground. These proposals do not always specify just one way of changing the common ground. They may suggest alternative ways of doing so, among which the responder is then invited to choose.

Formally, a proposition consists of one or more *possibilities*. Each possibility is a set of possible worlds—a set of *indices*, as we will call them—and embodies a possible way to change the common ground. If a proposition consists of two or more possibilities, it is *inquisitive*: it invites the other participants to respond in a way that will lead to a cooperative choice between the proposed alternatives. In this sense, inquisitive propositions raise an issue. They give direction to a dialogue. Purely informative non-inquisitive propositions do not invite other participants to choose between different alternatives. But still, they are proposals. They do not automatically establish a change of the common ground.

Thus, the notion of meaning in inquisitive semantics is directly related to the interactive process of exchanging information. Propositions, conceived of as proposals, give direction to this process. Changes of the common ground come about by mutual agreement among speech participants.

2.2 Support, Possibilities, and Propositions

We define an inquisitive semantics for a propositional language, which is based on a finite set of propositional variables, and has $\neg$, $\wedge$, $\vee$, and $\rightarrow$ as its basic logical operators. We add two non-standard operators: ! and ?. $!\varphi$ is defined as $\neg\neg\varphi$, and $?\varphi$ is defined as $\varphi \vee \neg\varphi$. $!\varphi$ is called the *non-inquisitive closure* of φ, and $?\varphi$ is called the *non-informative closure* of φ.

The basic ingredients for the semantics are *indices* and *states*. An index is a binary valuation for the atomic sentences in the language. A state is a non-empty set of indices. We use v as a variable ranging over indices, and σ, τ as variables ranging over states. The set of all indices is denoted by ω, and the set of all states is denoted by $\mathcal{S}$.

The proposition expressed by a sentence φ is defined indirectly, via the notion of *support* (just as, classically, the proposition expressed by a sentence is usually defined indirectly in terms of truth). We read $\sigma \models \varphi$ as *state* σ *supports* φ. Support is recursively defined as follows.

Definition 1 (Support)

1. $\sigma \models p$ *iff* $\forall v \in \sigma : v(p) = 1$
2. $\sigma \models \neg\varphi$ *iff* $\forall \tau \subseteq \sigma : \tau \not\models \varphi$
3. $\sigma \models \varphi \vee \psi$ *iff* $\sigma \models \varphi$ *or* $\sigma \models \psi$
4. $\sigma \models \varphi \wedge \psi$ *iff* $\sigma \models \varphi$ *and* $\sigma \models \psi$
5. $\sigma \models \varphi \rightarrow \psi$ *iff* $\forall \tau \subseteq \sigma$: *if* $\tau \models \varphi$ *then* $\tau \models \psi$

In terms of support, we define the possibilities for a sentence, and the proposition expressed by a sentence.

Definition 2 (Possibilities and Propositions)

1. *A* possibility *for* φ *is a maximal state supporting* φ, *that is, a state that supports* φ *and is not properly included in any other state supporting* φ.
2. *The* proposition *expressed by* φ, *denoted* $\lfloor\varphi\rfloor$, *is the set of possibilities for* φ.

We will illustrate the behavior of atomic sentences and the logical operators by means of the examples displayed in figure 1. In doing so, it will be useful to distinguish between *classical* and *inquisitive* sentences.

Definition 3 (Classical and Inquisitive Sentences)

1. φ *is* classical *iff* $\lfloor\varphi\rfloor$ *contains at most one possibility;*
2. φ *is* inquisitive *iff* $\lfloor\varphi\rfloor$ *contains at least two possibilities.*

Atoms. The proposition expressed by an atomic sentence p always consists of *exactly one* possibility: the possibility containing *all* indices that make p true. So atomic sentences are always classical.

Negation. The proposition expressed by $\neg\varphi$ always consists of *at most* one possibility. If there are indices that make φ false (classically speaking), then the unique possibility for $\neg\varphi$ consists of all such indices; if there are *no* indices that make φ false, then there is no possibility for $\neg\varphi$. In any case, negated sentences, like atomic sentences, are always classical.

Disjunction. Disjunctions are typically inquisitive. To determine the proposition expressed by a disjunction $\varphi \vee \psi$ we first collect all states that support φ or ψ. The maximal elements among these states are the possibilities for $\varphi \vee \psi$. Figures 1(a)–1(c) give some examples: a simple disjunction of two atomic sentences $p \vee q$, a polar question $?p$ (recall that $?p$ is defined as $p \vee \neg p$), and the disjunction of two polar questions $?p \vee ?q$.

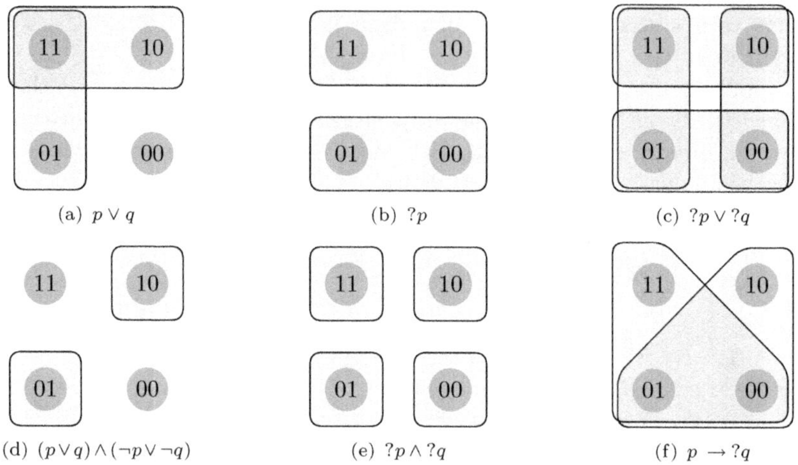

Fig. 1. Some examples of inquisitive sentences. It is assumed in these examples that there are only two proposition letters, p and q, and thus only four indices: 11 is the index where p and q are both true, 10 is the index where p is true and q is false, etc.

Conjunction. The proposition expressed by a conjunction $\varphi \wedge \psi$ consists of all maximal states supporting both φ and ψ. If φ and ψ are both classical, then conjunction amounts to intersection, just as in the classical setting. If φ and/or ψ are inquisitive, then the conjunction $\varphi \wedge \psi$ may be inquisitive as well. Figures 1(d) and 1(e) show what this amounts to for the conjunction of two disjunctions $(p \vee q) \wedge (\neg p \vee \neg q)$ and the conjunction of two polar questions $?p \wedge ?q$.

Implication. The proposition expressed by $\varphi \rightarrow \psi$ consists of all maximal states σ such that all substates of σ that support φ also support ψ. If the consequent ψ is classical, then $\varphi \rightarrow \psi$ behaves just as it does in the classical setting: in this case, $\lfloor \varphi \rightarrow \psi \rfloor$ consists of a single possibility, containing all indices that make ψ true or φ false. If the consequent ψ is inquisitive, then $\varphi \rightarrow \psi$ may be inquisitive as well. Figure 1(f) shows what this amounts to for a conditional question $p \rightarrow ?q$. There is much more to say about implication, but that would take us too far astray from the central concern of this paper (see, for instance, Groenendijk, 2009; Ciardelli and Roelofsen, 2009).

2.3 Truth-Sets and Excluded Possibilities

Besides the proposition expressed by a sentence φ it will also be useful to speak of the *truth-set* of φ, and of the possibility *excluded by* φ.

Definition 4 (Truth Sets). *The truth-set of φ, denoted by $|\varphi|$, is the set of indices where φ is classically true.*

In a classical setting, the truth-set of φ is simply the proposition expressed by φ. In the inquisitive setting, $|\varphi|$ is identical to the *union* of all the possibilities that make up the proposition expressed by φ. In both cases, $|\varphi|$ embodies the *informative content* of φ: someone who utters φ proposes to eliminate all indices that are not in $|\varphi|$ from the common ground.

Definition 5 (Excluded Possibility)

1. *If* $\omega - |\varphi| \neq \emptyset$, *then* $\omega - |\varphi|$ *is called the possibility excluded by* φ*;*
2. *If* $\omega - |\varphi| = \emptyset$, *then we say that* φ *does not exclude any possibility;*
3. *The (singleton- or empty) set of possibilities excluded by* φ *is denoted by* $\lceil\varphi\rceil$.

The semantics for $\neg$, ?, and ! can be stated in a transparent way in terms of exclusion (recall that $!\varphi$ was defined as $\neg\neg\varphi$ and $?\varphi$ as $\varphi \vee \neg\varphi$).

Fact 1 ($\neg$, ?, and ! in terms of exclusion)

1. $\lfloor\neg\varphi\rfloor = \lceil\varphi\rceil$
2. $\lfloor!\varphi\rfloor = \lceil\neg\varphi\rceil$
3. $\lfloor?\varphi\rfloor = \lfloor\varphi\rfloor \cup \lceil\varphi\rceil$

2.4 Questions, Assertions, and Hybrids

We already defined a sentence φ to be inquisitive just in case $\lfloor\varphi\rfloor$ contains at least two possibilities. Uttering an inquisitive sentence is one way of making a significant contribution to a conversation. The other way in which a significant contribution can be made is by being *informative*. A sentence φ is informative iff there is at least one possibility for φ, and also a possibility that φ excludes.

Definition 6 (Informative Sentences)
φ *is* informative *iff* $\lfloor\varphi\rfloor$ *and* $\lceil\varphi\rceil$ *both contain at least one possibility.*

In terms of whether a sentence is inquisitive and/or informative or not, we distinguish the following four semantic categories:

	informative	inquisitive
question	−	+
assertion	+	−
hybrid	+	+
insignificant	−	−

A *question* is inquisitive and not informative, an *assertion* is informative and not inquisitive, a *hybrid* sentence is both informative and inquisitive, and an *insignificant* sentence is neither informative nor inquisitive. Some examples are provided in figure 2.

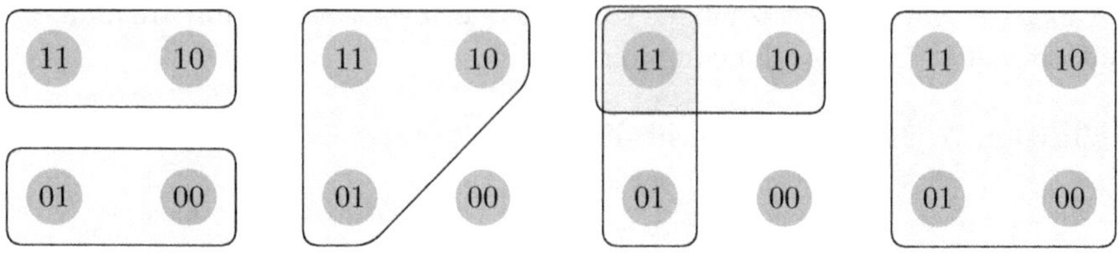

(a) Question (b) Assertion (c) Hybrid (d) Insignificant

Fig. 2. One example for each of the four semantic categories

2.5 Inquisitive Entailment

Classically, φ entails ψ iff the proposition expressed by φ is contained in the proposition expressed by ψ. In inquisitive semantics, every possibility for φ must be contained in some possibility for ψ.

Definition 7 (Entailment) $\varphi \models \psi$ *iff* $\forall \alpha \in \lfloor \varphi \rfloor : \exists \beta \in \lfloor \psi \rfloor : \alpha \subseteq \beta$

Entailment may also be formulated in terms of support rather than in terms of possibilities. This formulation is analogous to the classical formulation of entailment in terms of truth.

Fact 2 (Entailment in terms of support)
$\varphi \models \psi$ *iff every state that supports* φ *also supports* ψ.

If an assertion $!\varphi$ entails a question $?\psi$, then $!\varphi$ completely resolves the issue raised by $?\psi$. To some extent this means that $!\varphi \models ?\psi$ characterizes answerhood. We say to some extent since it only characterizes complete and not partial answerhood, and it is not very 'precise' in characterizing complete answerhood in that it allows for over-informative answers: if $!\varphi \models ?\psi$ and $!\chi \models !\varphi$, then also $!\chi \models ?\psi$.

For some questions, but not for all, we can characterize precise and partial answerhood in terms of entailment by saying that $!\varphi$ is an answer to $?\psi$ iff $?\psi \models ?!\varphi$. The intuition here is that $!\varphi$ is an answer to $?\psi$ just in case the polar question $?!\varphi$ behind $!\varphi$ is a subquestion of $?\psi$ (cf. Groenendijk, 1999; ten Cate and Shan, 2007).

This characterization gives correct results as long as we are dealing with questions whose possibilities are mutually exclusive. Such questions partition logical space. However, since in inquisitive semantics questions do not necessarily partition logical space, $?\psi \models ?!\varphi$ does not give us a general characterization of answerhood, and neither does $?\varphi \models ?\psi$ give us a general characterization of subquestionhood.

Problems arise as soon as we consider questions with *overlapping* possibilities. Conditional questions and alternative questions are questions of this kind. First, consider a conditional question $p \to ?q$ (*If Alf goes to the party, will Bea go as well?*). We certainly want $p \to q$ to count as an answer to this question,

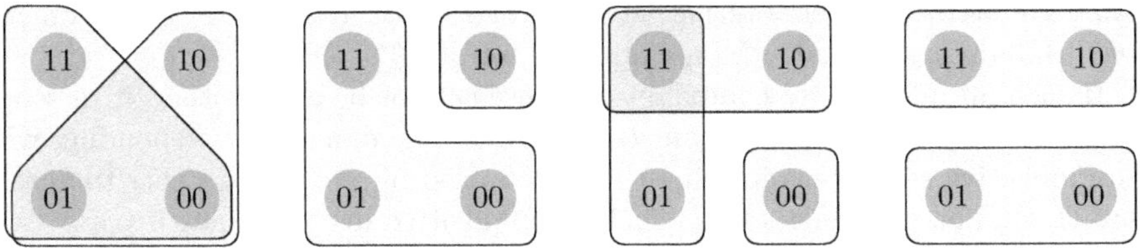

(a) $p \to ?q$ (b) $?(p \to q)$ (c) $?(p \vee q)$ (d) $?p$

Fig. 3. Questions with overlapping possibilities are problematic for characterizations of answerhood and subquestionhood in terms of entailment

but $p \to ?q \not\models ?(p \to q)$. This can easily be seen by inspecting the propositions expressed by $p \to ?q$ and $?(p \to q)$, depicted in figure 3(a) and 3(b). In fact, entailment obtains in the other direction in this case: $?(p \to q) \models p \to ?q$.

Similarly, we certainly want p to count as an answer to the alternative question $?(p \vee q)$ (*Does* ALF *or* BEA *go to the party?*). But $?(p \vee q) \not\models ?p$, as can be seen by comparing figure 3(c) and 3(d).

This does not mean that there is anything wrong with the entailment relation as such. It does what it should do: provide a characterization of meaning-inclusion. In particular, entailment between an assertion and a question means that the assertion fully resolves the issue raised by the question, and entailment between two questions $?\varphi$ and $?\psi$ means that the issue raised by $?\psi$ is fully resolved whenever the issue raised by $?\varphi$ is.

At the same time, given that entailment does not lead to a general notion of answerhood and subquestionhood, we surely are in need of a logical notion that does characterize these relations.

3 Compliance

The logical notion of *compliance* will judge whether a certain conversational move makes a significant contribution to resolving a given issue. Before stating the formal definition, however, let us first discuss the basic logico-pragmatical intuitions behind it.

Basic intuitions. Consider a situation where a sentence φ is a *response* to an *initiative* ψ. We are mainly interested in the case where the initiative ψ is inquisitive, and hence proposes several alternatives. In this case, we consider φ to be an optimally compliant response just in case it picks out exactly one of the alternatives proposed by ψ. Such an optimally compliant response is an assertion φ such that the unique possibility α for φ equals one of the possibilities for ψ: $\lfloor\varphi\rfloor = \{\alpha\}$ and $\alpha \in \lfloor\psi\rfloor$. Of course, the responder will not always be able to give such an optimally compliant response. It may still be possible in this case to give a compliant informative response, not by picking out *one* of the alternatives proposed by ψ, but by selecting some of them, and excluding others.

The informative content of such a response must correspond with the union of some but not all of the alternatives proposed by ψ. That is, $|\varphi|$ must coincide with the union of a proper non-empty subset of $\lfloor\psi\rfloor$.

If such an informative compliant response cannot be given either, it may still be possible to make a significant compliant move, namely by responding with an inquisitive sentence, replacing the issue raised by ψ with an easier to answer sub-issue. The rationale behind such an inquisitive move is that, if part of the original issue posed by ψ were resolved, it might become possible to subsequently resolve the remaining issue as well.

Summing up, there are basically two ways in which φ may be compliant with ψ:

(a) φ may partially *resolve* the issue raised by ψ;
(b) φ may *replace* the issue raised by ψ by an easier to answer sub-issue.

Combinations are also possible: φ may partially resolve the issue raised by ψ and at the same time replace the remaining issue with an easier to answer sub-issue.

Compliance and over-informative answers. Compliance does not allow for over-informative answers. For instance, p is a compliant response to $?p$, but $p \wedge q$ is not. More generally, if ψ is an inquisitive initiative, and φ and χ are two assertive responses such that the unique possibility for φ coincides with one of the possibilities for ψ, and the unique possibility for χ is properly included in the one for φ, then φ is regarded as optimally compliant, while χ is not consisered to be compliant at all, because it is over-informative.

The rationale behind this is that over-informative answers incur an unnecessary *risk* of being inconsistent with other participants' information states. If so, the proposal they express will be rejected. For instance, in the scenario just considered, φ and χ both have the potential to fully resolve the issue raised by ψ. However, χ is over-informative, and therefore unnecessarily runs a higher risk of being rejected. And if it is indeed rejected, the given issue remains unresolved.

These considerations are captured by the following definition:

Definition 8 (Compliance). *φ is compliant with ψ, $\varphi \propto \psi$, iff*

1. *every possibility in $\lfloor\varphi\rfloor$ is the union of a set of possibilities in $\lfloor\psi\rfloor$*
2. *every possibility in $\lfloor\psi\rfloor$ restricted to $|\varphi|$ is contained in a possibility in $\lfloor\varphi\rfloor$*

Here, the *restriction* of $\alpha \in \lfloor\psi\rfloor$ to $|\varphi|$ is defined to be the intersection $\alpha \cap |\varphi|$. To explain the workings of the definition, we will consider the case where ψ is an insignificant sentence, an assertion, a question, and a hybrid one by one.

If ψ is a contradiction, the first clause can only be met if φ is a contradiction as well. The second clause is trivially met in this case. Similarly, if ψ is a tautology, the first clause can only be met if φ is a tautology as well, and the second clause is also satisfied in this case. Thus, if ψ is insignificant, φ is compliant with ψ just in case φ and ψ are equivalent.

Fact 3. *If ψ is insignificant, then $\varphi \propto \psi$ iff $\lfloor\varphi\rfloor = \lfloor\psi\rfloor$.*

Next, consider the case where ψ is an assertion. Then the first clause says that every possibility for φ should coincide with the unique possibility for ψ. This can only be the case if φ is equivalent to ψ. In this case, the second clause is trivially met. Thus, the only way to compliantly respond to an assertion is to confirm it.

Fact 4. *If ψ is an assertion, then $\varphi \propto \psi$ iff $\lfloor\varphi\rfloor = \lfloor\psi\rfloor$.*

If ψ is a question and φ is an assertion, then the first clause in the definition of compliance requires that $|\varphi|$ coincides with the union of a set of possibilities for ψ. The second clause is trivially met in this case. Such an assertion provides information that is fully dedicated to partially resolving the issue raised by the question, and does not provide any information that is not directly related to the issue. Recall that at the end of section 2 we criticized the notion of entailment for not delivering a notion of 'precise' (partial) answerhood. This is precisely what compliance of assertions to questions characterizes.

Fact 5. *If ψ is a question and φ an assertion, then $\varphi \propto \psi$ iff $|\varphi|$ coincides with the union of a set of possibilities for ψ.*

If φ and ψ are both questions, then the first clause requires that φ is related to ψ in the sense that every complete answer to φ is at least a partial answer to ψ. In this case the second clause has work to do as well. However, since φ is assumed to be a question, and since questions are not informative, the second clause can be simplified in this case: the restriction of the possibilities for ψ to $|\varphi|$ does not have any effect, because $|\varphi| = \omega$. Hence,the second clause simply requires that every possibility for ψ is contained in a possibility for φ (i.e., that ψ entails φ). This constraint prevents φ from being more difficult to answer than ψ.

We illustrate this with an example. Consider the case where $\psi \equiv ?p \vee ?q$ and $\varphi \equiv ?p$. The propositions expressed by these sentences are depicted in figure 4.

Intuitively, $?p \vee ?q$ is a *choice question*. To resolve it, one may either provide an answer to the question $?p$ or to the question $?q$. Thus, there are four possibilities, each corresponding to an optimally compliant response: p, $\neg p$, q and $\neg q$. The question $?p$ is more demanding: there are only two possibilities and thus only two optimally compliant responses, p and $\neg p$. Hence, $?p$ is more difficult to answer than $?p \vee ?q$, and should therefore not count as compliant with it. This is

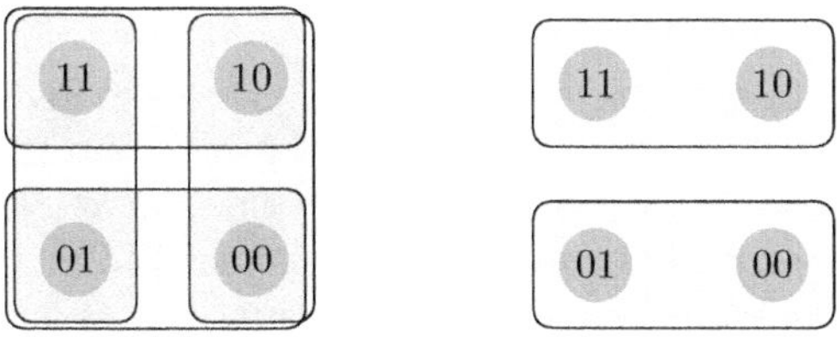

(a) $?p \vee ?q$ (b) $?p$

Fig. 4. Two non-compliant questions

not taken care of by the first clause in the definition of compliance, since every possibility for $?p$ is also a possibility for $?p \vee ?q$. So the second clause is essential in this case: it says that $?p$ is not compliant with $?p \vee ?q$ because two of the possibilities for $?p \vee ?q$ are not contained in any possibility for $?p$. The fact that these possibilities are, as it were, 'ignored' by $?p$ is the reason that $?p$ is more difficult to answer than $?p \vee ?q$.

Recall that at the end of section 2, we criticized the notion of entailment for not delivering a satisfactory notion of subquestionhood. The difference with compliance, which does give the right characterization, lies in the first clause of the definition, which requires that the two questions are *related.* As we saw above, entailment only covers (the simplified version of) the second clause.

Fact 6. *If both ψ and φ are questions, then $\varphi \propto \psi$ iff*

1. *every possibility in $\lfloor\varphi\rfloor$ is the union of a set of possibilities in $\lfloor\psi\rfloor$*
2. *every possibility in $\lfloor\psi\rfloor$ is contained in a possibility in $\lfloor\varphi\rfloor$*

The second clause in the definition of compliance only plays a role in case both φ and ψ are inquisitive. Moreover, the restriction of the possibilities for ψ to $|\varphi|$ can only play a role if $|\varphi| \subset |\psi|$, which is possible only if φ is informative. Thus, the second clause can only play a role in its unsimplified form if φ is both inquisitive and informative, i.e., hybrid. If φ is hybrid, just as when φ is a question, the second clause forbids that a possibility for ψ is ignored by φ. But now it also applies to cases where a possibility for ψ is partly excluded by φ. The part that remains should then be fully included in one of the possibilities for φ.

As an example where this condition applies, consider $p \vee q$ as a response to $p \vee q \vee r$. One of the possibilities for $p \vee q \vee r$, namely $|r|$, is ignored by $p \vee q$: the restriction of $|r|$ to $|p \vee q|$ is not contained in any possibility for $p \vee q$. Again, this reflects the fact that the issue raised by $p \vee q$ is more difficult to resolve than the issue raised by $p \vee q \vee r$.

A general characterization of what the second clause says, then, is that φ may only remove possibilities for ψ by providing information. A possibility for ψ must either be excluded altogether, or it must be preserved: its restriction to $|\varphi|$ must be contained in some possibility for φ.

4 Homogeneity: Say More, Ask Less!

There may be several possible compliant responses to a given initiative. Among these compliant responses, some may be preferable over others. The main point of this section—as is foretold by its title—is that there is a general preference for *more informative*, and *less inquisitive* responses. To make this more precise, let us introduce comparative notions of informativeness and inquisitiveness. In order to do so, we first need to relativize the semantic notions defined in section 2 to information states.

Definition 9 (Relative Semantic Notions)

1. *A* possibility for φ in σ *is a maximal substate of* σ *supporting* φ.
2. *The* proposition expressed by φ in σ, *denoted by* $\sigma\lfloor\varphi\rfloor$, *is the set of possibilities for* φ *in* σ.
3. *We say that* φ excludes a possibility in σ *iff the union of all the possibilities for* φ *in* σ *is not identical to* σ *itself.*
4. φ *is* inquisitive in σ *iff there are at least two possibilities for* φ *in* σ*;*
5. φ *is* acceptable in σ *iff there is at least one possibility for* φ *in* σ*;*
6. φ *is* eliminative in σ *iff* φ *excludes a possibility in* σ*;*
7. φ *is* informative in σ *iff* φ *is both acceptable in* σ *and eliminative in* σ.

Definition 10 (Comparative Informativeness and Inquisitiveness)

1. φ is at least as informative as ψ *iff in every state where* ψ *is eliminative,* φ *is eliminative as well.*
2. φ is at most as inquisitive as ψ *iff in every state where* ψ *is not inquisitive,* φ *is not inquisitive either.*

Note that comparative informativeness is defined in terms of eliminativity. If it were formulated in terms of informativity, it would give very counter-intuitive results. Suppose that we defined φ to be at least as informative as ψ iff in every state where ψ is informative, φ is informative as well. Then, for instance, $p \wedge q$ would not count as more informative than p. To see this consider the state $|\neg q|$. In this state, p is informative, but $p \wedge q$ is *not*, because it is *unacceptable* in $|\neg q|$. More generally, for any non-tautological sentence χ, it would be impossible to find a formula that is more informative than χ. This is clearly undesirable. Thus, in order to measure comparative informativeness, the acceptability aspect of informativeness must be left out of consideration—the only relevant feature is eliminativity.

Now let us motivate the general preference for more informative and less inquisitive responses to a given initiative. In each case, we will provide a general argument, and a concrete example.

Say More! Consider an inquisitive initiative ψ and two compliant assertive responses φ and χ, such that φ is more informative than χ. This means that φ rules out more of the possibilities proposed by ψ than χ does. In this sense, φ more fully resolves the issue raised by ψ, and thus makes a more substantial contribution to enhancing the common ground than χ does. Therefore, φ is preferred over χ.

To illustrate this with a concrete example, consider a conversation between two people, A and B. Suppose A utters $?p \wedge ?q$. This sentence expresses a proposition consisting of four possibilities (see figure 1). Now, consider q and $p \rightarrow q$, which are both compliant responses to A's initiative. q is more informative than $p \rightarrow q$. In particular, q rules out two of the possibilities proposed by $?p \wedge ?q$, while $p \rightarrow q$ only rules out one of these possibilities. Therefore, q is preferred.

However, q is not yet optimal. An even more informative compliant response is $p \wedge q$. This response picks out exactly one of the possibilities proposed by $?p \wedge ?q$, and thus fully resolves the issue. In general, one compliant response is preferred over another if it more fully resolves the given issue.

Ask Less! Now consider an initiative ψ and an *inquisitive* compliant response φ. In this case, φ raises a sub-issue, which addresses the original issue in an indirect way. The hope is that the sub-issue may be resolved first, and that, subsequently, there will be a better chance of resolving the original issue as well. Now, this strategy will only work if it is indeed possible to resolve the sub-issue first. And this is more likely to be the case if φ is *less inquisitive.* This is why less inquisitive responses are generally preferred over more inquisitive responses.

To illustrate this, consider again the example sketched above. As before, suppose that A raises an issue by uttering $?p \wedge ?q$ (see figure 1). But now suppose that B is not able to resolve this issue directly. Then he may try to resolve it indirectly by raising a sub-issue. Consider the following two sentences that B may utter in this situation: $?q$ and $p \rightarrow ?q$ (see again figure 1). Now, it is very unlikely that A will have an answer to $?q$, given that he has just asked $?p \wedge ?q$ himself. On the other hand, it is not so unlikely that A will have an answer to $p \rightarrow ?q$. This question is weaker than $?q$, it merely asks whether or not p and q are related in a certain way. Thus, it is much more advisable for B to ask $p \rightarrow ?q$ than to ask $?q$. Both $?q$ and $p \rightarrow ?q$ are compliant with the original question. But $p \rightarrow ?q$ is preferred because it is less inquisitive.

These considerations lead to the following definitions:

Definition 11 (Homogeneity)

> φ is at least as homogeneous as χ, $\varphi \succeq \chi$ *iff* φ *is at least as informative and at most as inquisitive as* χ.

Definition 12 (Comparative Compliance)

> φ *is a* more compliant *response to* ψ *than* χ *iff* φ *and* χ *are both compliant responses to* ψ*, and* φ *is more homogeneous than* χ.

Finally, the following fact characterizes most and least compliant responses.

Fact 7 (Ultimate Compliance)

1. φ *is a least compliant response to* ψ *iff* φ *is equivalent to* ψ.
2. φ *is a most compliant response to* ψ *iff there is a single possibility* α *for* φ*, and* α *is a possibility for* ψ *as well.*
3. *If* ψ *is a question,* φ *is a most compliant non-informative response to* ψ *iff* φ *is a polar sub-question of* ψ.

5 Conclusion

We have specified and given motivation for the logical notion of compliance, which determines whether a sentence makes a contribution to resolving a given issue. In particular, this notion yields a characterization of answerhood and subquestionhood. And the additional notion of homogeneity captures certain quantitative preferences among compliant responses.

To be sure, we have abstracted away from certain issues that should certainly be considered in a more comprehensive analysis. For instance, whether or not a sentence makes a contribution to a given issue partly depends on the information that is already available. For now, we hope to have made a convincing initial case for the logical and linguistic interest of the notion of compliance.

References

Ciardelli, I., Roelofsen, F.:Inquisitive Logic. Journal of Philosophical Logic, published online (July 16, 2010)

Groenendijk, J.: The logic of interrogation. In: Matthews, T., Strolovitch, D. (eds.) Semantics and Linguistic Theory, pp. 109–126. Cornell University Press, Ithica (1999)

Groenendijk, J.: Inquisitive semantics: Two possibilities for disjunction. In: Bosch, P., Gabelaia, D., Lang, J. (eds.) Seventh International Tbilisi Symposium on Language, Logic, and Computation. Springer, Heidelberg (2008)

Groenendijk, J.: Inquisitive semantics: Questions, assertions, and hybrids, Amsterdam (2009), `http://www.illc.uva.nl/inquisitive-semantics` (manuscript)

Groenendijk, J., Stokhof, M.: Studies on the Semantics of Questions and the Pragmatics of Answers. Ph.D. thesis, University of Amsterdam (1984)

Mascarenhas, S.: Inquisitive semantics and logic. MSC Thesis, ILLC, University of Amsterdam (2009)

Roberts, C.: Information structure in discourse. In: Yoon, J., Kathol, A. (eds.) OSU Working Papers in Linguistics, vol. 49, pp. 91–136. Ohio State University (1996)

Stalnaker, R.: Assertion. Syntax and Semantics 9, 315–332 (1978)

ten Cate, B., Shan, K.: Axiomatizing Groenendijk's logic of interrogation. In: Aloni, M., Butler, A., Dekker, P. (eds.) Questions in Dynamic Semantics, pp. 63–82. Elsevier, Amsterdam (2007)

The Calculus of Responsibility and Commitment

Carl Pollard

INRIA-Lorraine and Ohio State University

1 Introduction

Ever since Montague (1974 [1970]) laid the foundations for formally precise analysis of natural language (hereafter NL) semantics in the late 1960's, the typed lambda calculus (hereafter TLC) and certain of its extensions[1] have been the linguists' tool of choice for representing the meanings of NL expressions. But starting around the turn of the millenium, motivated by a range of linguistic phenomena collectively known as **covert movement** phenomena[2], logical grammarians of various persuasions have proposed the use of other semantic term calculi that embody, directly or indirectly, some notion or other of **continuation**.

In this paper, I will do the same thing. My justification for stepping into the fray, in spite of my distinctly amateur standing vis-à-vis the mathematics and computer science of continuations, is that the proposals I have seen so far seem, for various reasons, not yet able to compete with conventional Montague-style compositional semantics in the linguistic marketplace. Some of the proposals appear to be incompletely specified; others require technical knowledge of mathematics or computer science that almost no linguists control (or even have straightforward access to); and in others, the sheer complexity seems incommensurate with the difficulty of the problems to be solved. At the heart of my proposal is a term calculus called RC which is suggested by way of a replacement for TLC as a notational system for NL meanings. It is my intention that RC be as easy for linguists to learn and use as TLC is, while at the same time doing the semantic heavy lifting performed by existing systems of continuized semantics.

This paper is highly programmatic in nature. For more of the linguistic motivation and analysis of linguistic examples, the reader is referred to the companion piece (Pollard in press).

---

[1] Specifically, variants of Church's (1940) and Henkin's (1950) simple theory of types, especially Montague's own Intensional Logic (IL) and Gallin's (1975) Ty2.

[2] So-called because they are analyzed by transformational generative grammarians via tree operations on syntactic representations which are reflected in the **logical form (LF)** of the sentence being analyzed, but which take place too late in the derivation to be reflected in the **phonetic form (PF)**. By no means does the use of the term 'covert movement' imply advocacy of the technology of transformational grammar, such as arboreal structural representations of linguistic expressions (not to mention derivations upon such representations made up of structural operations that delete, copy, or move subtrees.

Lecomte and S. Tronçon (Eds.): PRELUDE 2010, LNAI 6505, pp. 174–187, 2011.

2 Background

The so-called covert movement phenomena involve a linguistic expression that seems, pretheoretically speaking, to have semantic scope, but to occupy a syntactic position appropriate not for an operator with such a scope, but rather, for a variable bound by such an operator. In the parlance of transformational grammar, the operator remains, in overt syntax, **in situ**, and moves to its scope position only covertly, leaving in its place a variable which it binds at the level of LF. Examples of in situ operators include (but are by no means limited to) the following:

(1) **Quantificational Noun Phrases**
Kim thinks [**everyone** walks].

Quantifer scope ambiguity: the QNP **everyone** can take scope in either the root clause or the embedded clause.

(2) **In-situ interrogative expressions (wh-in situ)**
Who asked [who ordered **what**]?

Ambiguity of wh-in-situ construal: the in-situ wh-expression **what** can scope in either the root question or the embedded question ('Baker's ambiguity').

(3) **Pied Piping**

a. [To **whose** parents]$_i$ were you speaking t_i?
b. These are the reports [the height of the lettering on the covers of **which**]$_i$ the government prescribes t_i.

The 'pied piped' (interrogative or relative) expression is *in situ* relative to a larger expression which undergoes 'overt movement'.

(4) **Comparative Word (more/less) or Morpheme (-er)**
Russell thought [the yacht was long**er**] than it was.

Ambiguity of comparative operator construal: the comparative morpheme **-er** can scope in either the root question (how long Russell thought the yacht was exceeded the actual length) or in the embedded clause (Russell stupidly thought the yacht's length was self-exceeding).

(5) **Focus**
Mary thinks that **John**[H*] will vote for McCain.

The focal or 'A' [H*] pitch accent on **John** somehow conveys that there is at least one contextually salient individual different from John who Mary does *not* think will vote for McCain.

(6) **In-situ Contrastive Topic**
He thinks she likes **bagels**[L+H*].

The contrastive-topic or 'B' [L+H*] pitch accent on **bagels** somehow conveys that whether he thinks she likes certain things, including bagels, is at issue, and that the answers may vary.

(7) **Phrasal-Comparative Associates**

a. **The Pope** goes to Minneapolis more often than St. Paul. (St. Paul = the saint)

b. The Pope goes to **Minneapolis** more often than St. Paul. (St. Paul = the city)

Roughly speaking, the pitch-accented phrase in the main clause (the **associate**) and the complement phrase to *than* (the **remnant**) are plugged into the same context and then the least upper bounds of the two resulting sets of degrees compared.

(8) **Superlative Associates**

a. **Kim** likes Sandy the most. (Kim likes Sandy more than anyone else likes Sandy.)

b. Kim likes **Sandy** the most. (Kim likes Sandy more than Kim likes anyone else.)

c. **Kim** likes **Sandy** the most. (Kim likes Sandy more than anyone else likes anyone else.)

A superlative sentence asserts that the interpretation of the (possibly discontinuous) associate is where a certain degree-valued function assumes its maximum value.

If we call the linguistic material surrounding the in situ operator its **linguistic context** and that part of the linguistic context which expresses the operator's scope the **(delimited) continuation**[3] of the operator, then it is easy to see why Barker (2002), who pioneered continuation semantics for NL, asserted what he called the Continuation Hypothesis:

(9) **Barker 2002**

a. **The Continuation Hypothesis:** Some linguistic expressions (in particular, QNPs) have denotations that manipulate their own continuations.

b. Barker shows how to transform a CFG with conventional Montague-style compositional semantics into one with 'continuized semantics', by analogy with Fischer's (1972)/Plotkin's (1975) **call-by-value CPS transform**.

[3] Here, the sense of 'delimited' is that the scope of the operator need not be the entire linguistic context (the root clause).

c. In terms, variables with overbars are called 'continuation variables' and certain implicative types with result type t are called 'continuation types'.

d. *Formally*, these are ordinary variables and ordinary types; the semantic calculus is embedded into ordinary (intuitionistic) TLC.

(10) **de Groote 2001**

a. de Groote uses a nonconfluent calculus with an involutive negation (similar in some respects to Parigot's (2000) **symmetric $\lambda\mu$-calculus**), with Montague's type t as $\perp$, to analyze one example of quantifier scope ambiguity.

b. Unlike Barker's CPS approach, this one (like those of Shan and Bernardi/Moortgat mentioned below, is in 'direct-style'.

c. Though it is claimed that $\lambda\mu$-calculus "allows Cooper's (1983) storage to be given a type-logical foundation", it is not clear how to generalize this approach to operators whose scope has type different from t, or whose result type differs from the scope type.

(11) **Shan 2004**

a. Shan pioneered the use of **delimited continuations** (Felleisen 1988) in NL.

b. The term calculus ('logical metalanguage') mixes syntactic and semantic constructs (for example it has directional applications and abstractions).

c. The ternary type constructor A_B^C provides the types for operators that 'scope over a B to bind an A-variable, resulting in a C.'

d. This constructor is a semantic analog of Moortgat's (1996) q-constructor.

e. A_B^C is mapped by the CPS transform to $(A \to B) \to C$.

f. Besides familiar TLC and Lambek-calculus constructs, the language employs a **hierarchy of control operators** (generalized shift and reset supercripted with 'strength levels', Danvy and Filinsky 1990).

(12) **Bernardi and Moortgat 2007**

a. Curien and Herbelin's (2000) **$\bar{\lambda}\mu\tilde{\mu}$-calculus** "is to sequent calculus as [the original, not the symmetric] $\lambda\mu$-calculus [Parigot 1992] is to natural deduction."

b. Bernardi and Moortgat develop a linear, directional variant of $\bar{\lambda}\mu\tilde{\mu}$-calculus to provide a semantic interpretation for **symmetric categorial grammars** based on the Lambek-Grishin calculus (Moortgat 2007).

c. Unlike other systems of continuation semantics (and predecessors such as Cooper 1983 and Hendriks 1993), but like mainstream categorial grammar, the Bernardi-Moortgat semantics is **functional** in the sense that interpretation is a function from syntactic proofs to meanings.

3 Cooper Storage and Retrieval

Unlike the approaches mentioned in the previous section, the inspiration for the RC calculus comes not from computer science but from linguistics, more specifically the storage-and-retrieval approach to scope developed by Robin Cooper (1975, 1983). The basic idea is easy to describe in a way that a first-year linguistic graduate student with a basic understanding of TLC can grasp almost immediately.

(13) **Cooper Storage and Retrieval (Intuitively)**

a. Suppose that while semantically interpreting a syntactic derivation bottom up, you encounter an in-situ operator with semantics a of semantic type $(A \rightarrow B) \rightarrow C$. Then (**storage**) you can replace a by a variable $x : A$ and place the pair (a, x) in the store.

b. Stores 'percolate upward' through derivations.

c. If you get to a node in the derivation with semantics $b : B$, then (**retrieval**) you can remove (a, x) from the store and replace $b : B$ by $a(\lambda_x b) : C$.

d. In case the in-situ operator is a QNP, $A = \text{e}$, and $B = C = \text{t}$.

(14) **Sociology of Cooper Storage and Retrieval**

a. Cooper didn't describe it this way. He expressed it all in terms of Montagovian model theory.

b. Cooper was at pains to avoid using λ-calculus, to make it clear he was not advocating some form of LF.

c. Even though this technology makes syntax much simpler (e.g. QNPs are just NPs), the resulting nonfunctionality of semantic interpretation was widely viewed with alarm.

d. However it was embraced in some quarters (e.g. Bach and Partee 1980, and in HPSG).

e. Categorial grammarians tend to describe this technology as 'noncompositional', 'baroque', 'ad hoc', 'a mess', 'as bad as covert movement', etc. (some of the more charitable characterizations).

f. The schemes of continuized semantics mentioned above are presented as major improvements on Cooper storage and retrieval.

(15) **Why RC Calculus?**

a. In spite of its reputation, Cooper's basic approach to scoping in situ operators is simple, illuminating, and easy to grasp.

b. But Cooper's insistence on a model-theoretic presentation made storage look more complicated than it really was.

c. RC calculus is designed to render the intuitions underlying Cooper storage more immediately graspable, via a purely syntactic presentation.

(16) **Toward RC Calculus**

a. RC is a term calculus in the labelled Gentzen-sequent style of natural deduction.
b. As far as I have been able to discern so far, RC can do all the linguistic work that has been done so far with continuized semantics.
c. RC can be directly semantically interpreted, but the easiest way is to transform RC meaning terms into TLC.
d. This transform is the RC analog of CPS transforms, but much simpler.
e. Morover RC meaning terms look familiar to people used to LF.

4 RC Types, Terms, and Commitments

Like TLC, RC has types, terms, and typing judgments. Also like TLC, the variable environment of an RC typing judgment is just a set of variable/type pairs to the left of the turnstile, which we will call the **variable context**; but an RC variable environment additionally has a **Cooper store**, written to the right and demarcated by a **co-turnstile** $\dashv$:

(17) **Format for RC Typing Judgments**
$\Gamma \vdash a : A \dashv \Delta$

The Cooper store is also called the **variable co-context**; the 'co-' here is mnemonic not only for 'Cooper'; but also for 'Commitment' (for reasons to be explained soon), for 'Covert Movement', and for 'Continuation' (since the operators stored in them will scope over their own continuations). Thus a judgment like (17) is read 'the term a is assigned the type A in the context Γ and the co-context Δ.'

(18) **RC Types**

a. There are some **basic** types.
b. If A and B are types, then $A \rightarrow B$ is an **implicative** type with **antecedent** type A and **consequent** type B.
c. If A, B, and C are types, then $\mathrm{O}[A, B, C]$, usually abbreviated (following Shan 2004) to A_B^C, is an **operator** type with **binding** type A, **continuation** type (or **scope** type) B, and **result** type C.

(19) **Basic Types**
Here, for expository simplicity, just Montague's extensional types e and t. But to do serious semantics, we really need to work in an intensional (or hyperintensional) semantic type theory.

(20) **Implicative Types**
No surprises here.

(21) **Operator Types**

a. These will be the semantic types for expressions which would be analyzed in TG as undergoing $\bar{\text{A}}$-movement (either overt or covert).

b. The O-constructor is essentially like Moortgat's (1996) q-constructor, with the crucial difference that it belongs to the *semantic* logic, *not* the syntactic one.

c. Thus, for example, while for Moortgat (1996) a QNP would have syntactic category $\mathrm{q[NP, S, S]}$ and semantic type $(\mathrm{e} \to \mathrm{t}) \to \mathrm{t}$, in RC it has syntactic type (simply) NP and semantic type $\mathrm{e}_{\mathrm{t}}^{\mathrm{t}}$[4]

d. Operator types are similar to Shan's (2004) **impure** types.

(22) **RC Terms**

a. There is a denumerable infinity of **variables** of each type.

b. There can be finitely many **basic constants** of each type.

c. There are **functional** terms of the form $(f\ a)$, where f and a are terms.

d. There are **binding** terms of the form $(a_x b)$ where a and b are terms and x is a variable.

e. But there is no λ binder.

(23) **What goes into the Cooper Store?**

a. The Cooper stores (co-contexts) will contain operators to be scoped, each paired with the variable that it will eventually bind.

b. We call such stored pairs **commitments**, and write them in the form a_x, where the type of x is the binding type of a.

c. Then we call x a **committed** variable, and say that a is **committed** to bind x.

d. By contrast, the variables in the (left-of-turnstile) context are called **uncommited** variables.

We can now give the rules of RC.

5 RC Rules

(24) **Schema H (Hypotheses)**

$x : A \vdash x : A \dashv$ (x fresh)

[4] Actually QNPs have to be polymorphically typed as $\mathrm{e}_{\mathrm{t}_\sigma}^{\mathrm{t}_\sigma}$ where σ ranges over strings of types and $A_\epsilon =_{\mathrm{def}} A$, $A_{B\sigma} =_{\mathrm{def}} B \to A_\sigma$. This is necessary to account for the fact that QNPs can scope not only over propositions, but also over propositional functions (e.g. ones of type $\mathrm{e} \to \mathrm{t}$, as in *Kim tried [to find a unicorn)* or *Mia knows every [owner of a hash bar])*. I am grateful to Patrick Blackburn and Scott Martin for alerting me to this issue. It does not arise in HPSG because VPs and $\bar{\text{N}}$s have propositional semantics (since arguments are incorporated into predicates by structure-sharing, not by Modus Ponens).

a. This is the usual ND schema for positing hypotheses.
b. It is the sole mechanism provided for introducing uncommitted variables into a semantic derivation.
c. We presuppose a syntax-semantics interface that recursively specifies a set of syntactic-semantic **derivation pairs** (Pollard in press). An uncommitted semantic variable will always be paired with a trace (syntactic variable) [5].

(25) **Schema A (Nonlogical Axioms)**

$\vdash c : A \dashv$ (c a basic constant of type A)

a. The basic constants notate lexical meanings, not just of words but also of bound morphemes that function as basic syntactic elements, such as argument clitics, phrasal affixes, sentence particles, etc. (some of which might have intonational or even empty phonological realizations).
b. The rules of the syntax-semantics interface that pair basic syntactic entities with their meanings are the **lexicon**.

(26) **Schema M (Modus Ponens)**

If $\Gamma \vdash f : A \rightarrow B \dashv \Delta$ and $\Gamma' \vdash a : A \dashv \Delta'$
then $\Gamma; \Gamma' \vdash (f\ a) : B \dashv \Delta; \Delta'$

a. This is the usual ND Modus Ponens, except that co-contexts as well as contexts have to propagated from premisses to conclusions (cf. (13b)).
b. Semicolons in (co-)contexts represent set union (necessarily disjoint, since variables are always posited fresh).
c. The syntax-semantics interface will pair instances of M with syntactic **merges**.

(27) **Schema C (Commitment)**

If $\Gamma \vdash a : A_B^C \dashv \Delta$ then $\Gamma \vdash x : A \dashv a_x : A_B^C; \Delta$ (x fresh)

a. This is a straightforward ND formulation of Cooper storage (13a).
b. It generalizes Carpenter's (1997) ND Introduction rule for Moortgat's (1988) $\Uparrow$ (essentially the special case of q where the scope type and the result type are the same), but in the semantics, **not** in the syntax.

(28) **More about Schema C**

a. The type of a committed variable always matches the binding type of the operator it is committed to.

[5] This is in keeping with an overall **parallel** grammar architecture, as opposed to the **cascaded** achitecture of mainstream generative grammar, which requires that, somehow or other, traces be replaced by 'logical variables' during the derivation from overt syntax to LF.

b. The syntax-semantics interface will guarantee that when an operator gets committed in the semantic derivation, **no corresponding syntactic change takes place**.
c. This is one of two reasons why the relation between syntax and semantics is **not** a function.
d. In this respect, our proposed architecture for the syntax-semantics interface resembles Cooper 1975/1983, Hendriks 1993, and HPSG, as well as most of the continuized-semantics proposals.
e. This sets it apart from mainstream categorial grammar, as well as the Bernardi-Moortgat functional style of continuized semantics.

(29) **Schema R (Responsibility)**

If $\Gamma \vdash b : B \dashv a_x : A_B^C; \Delta$ then $\Gamma \vdash (a_x b) : C \dashv \Delta$
(x free in b but not in Δ)[6]

a. This is a straightforward ND formulation of Cooper retrieval (13c).
b. It generalizes Carpenter's (1997) Elimination rule for Moortgat's $\Uparrow$, but, again, in the semantics, **not** in the syntax.
c. It is called Responsibility because it is about fulfilling commitments.
d. As with Commitment, the syntax-semantics interface will ensure that instances of Responsibility correspond to no syntactic change.
e. This is the other reason why the relation between syntax and semantics is **not** a function.

(30) **Schema O (Outside Assistance)**

If $\Gamma \vdash a : A_B^C \dashv \Delta$ and $x : A, \Gamma' \vdash b : B \dashv \Delta'$ then $\Gamma; \Gamma' \vdash (a_x b) : C \dashv \Delta, \Delta'$

The schema is so-called because if you cannot discharge your responsibilities (to make sure all variables eventually get bound) on your own (as the in-situ operators within you fulfil their commitments), then you have to get help from somewhere else.

(31) **More about Schema O**

a. O is also mnemonic for Overt Movement, because in TG the expression whose meaning is a would be analyzed as having moved (in overt syntax) from the position occupied by the trace (which somehow gets converted to x in between overt syntax and LF).
b. Note the strong similarity of the conclusions in Schemas R and O: in both cases the operator a binds the x in b to produce a C.
c. The difference is that in the former (in situ binding) case, b is a's continuation.

6 This side condition and the one for Schema C express the **linearity** of co-contexts. They rule out illicit retrievals of the kind that motivated Keller's (1988) nested stores (see Blackburn and Bos pp. 122-125 for discussion).

d. The idea for this schema was triggered by Guy Perrier's recent observation (p.c.) that the Topicalization Rule in Pollard 2007 was an elimination rule for the movement flavor of implication (SLASH, similar to the $\uparrow$ connective of Bach 1981 and Moortgat 1988).

(32) **Schema O vs. Anoun and Lecomte's $\multimap$ IE**

a. The $\multimap$ IE schema and Schema O have the same philosophy: to 'encapsulate' implication introductions within rules that immediately eliminate the implication.

b. But $\multimap$ IE has $(A \multimap B) \multimap C)$ in place of A_B^C, and $a(\lambda_x b)$ in place of $(a_x b)$. A problem with that formulation is that not all expressions with types of the form $(A \multimap B) \multimap C$ can undergo 'overt movement'.

c. In $\multimap$ IE, *controlled* hypotheses constrain extraction paths analogously to the use of functional uncertainty in LFG.

d. $\multimap$ IE is inspired by Vermaat's (1999) multimodal CG reformulation of Stabler's (1999) computational embodiment of Chomsky's (1995) MP.

e. Whereas Schema O is inspired by Gazdar's (1979) linking schemata for topicalization, *wh*-relatives, and *wh*-questions (his (28), (51), and (57), via GPSG and HPSG.

f. Unfortunately the feature-structural encoding of HPSG obscured the fact that it was essentially (albeit unknowingly) a natural-deduction system.

6 Conclusion

There are still a few loose ends to tie up. For one thing, so far we have not analyzed even one example. Since this paper is already getting overlong for a workshop paper, we offer just the obligatory QNP scope-ambiguity example and refer to Pollard in press for other examples (and for an explication of the Convergent Grammar (CVG) framework that the semantic approach sketched here is intended to work with).

(33) **Obligatory QNP Scope Ambiguity Example**

a. Kim thinks everyone walks.

b. CVG syntactic analysis:
$\vdash (^{\mathrm{S}}\ \mathsf{Kim}\ (\mathsf{thinks}\ (^{\mathrm{S}}\ \mathsf{everyone\ walks})\ ^{\mathrm{C}})) : \mathrm{S}$

All the subterms here are instances of the CVG syntactic schemas SM (Subject Merge) or CM (Complement Merge). The absence of a co-turnstile here is not a typo: syntactic derivations do not have co-contexts!

c. Narrow-scope semantic analysis:
$\vdash ((\mathsf{think'}\ (\mathsf{everyone'}_x(\mathsf{walk'}\ x)))\ \mathsf{Kim'}) : \mathsf{t} \dashv$

d. Wide-scope semantic analysis:
$\vdash \mathsf{everyone'}_x\ ((\mathsf{think'}\ (\mathsf{walk'}\ x))\ \mathsf{Kim'}) : \mathsf{t} \dashv$

Second, what do these RC terms really mean? The easiest way to explain this is to translate them into TLC, since everybody knows how to semantically interpet that. Fortunately, the translation from RC to TLC is a considerably simpler than the CPS transforms of (say) the $\bar{\lambda}\mu\tilde{\mu}$-calculus:

(34) **From RC to TLC**

a. First make sure the term is responsible (the Cooper store is empty), because commitments have no translation into TLC.
b. Replace every operator type A_B^C by $(A \to B) \to C)$, and every binding subterm $(a_x b)$ by $(a\ \lambda_x b)$.[7]
c. By convention we write application terms as $(f\ a)$ in RC and as $f(a)$ in TLC, just to make it easy to tell at a glance which calculus the term belongs to.

For example, the TLC translations of the RC terms in (33) are as expected:

(35) **TLC Terms for Scope Ambiguity Example**

a. Narrow-scope semantic analysis:
$\mathsf{think'}(\mathsf{everyone'}(\lambda_x \mathsf{walk'}(x)))(\mathsf{Kim'}) : \mathsf{t}$
b. Wide-scope semantic analysis:
$\mathsf{everyone'}(\lambda_x \mathsf{think'}(\mathsf{walk'}(x))(\mathsf{Kim'})) : \mathsf{t}$

Finally, since (in case it is still not obvious) the name RC is a (minimally) veiled reference to Robin Cooper, I want to emphasize that when he first developed his theory of storage and retrieval, Cooper by no means advocated a term-calculus formulation of the theory, quite the contrary in fact. So it's anybody's guess whether he will mind my naming a proof-theoretic embodiment of his theory after him. But since he has been working in Martin-Löf type theory for a number of years now, I like to think tht perhaps he won't mind too much.

Ackowledgments

For helpful discussion and comments on earlier stages of this work, I am grateful to Chris Barker, Patrick Blackburn, Wojciech Buszkowski, Robin Cooper, David Dowty, Philippe de Groote, Ruth Kempson, Brad Kolb, Oleg Kiselyov, Yusuke Kubota, Alain Lecomte, Tim Leffel, Jim Lambek, Scott Martin, Michael Moortgat, Glyn Morrill, Reinhard Muskens, Guy Perrier, Andy Plummer, Ken Shan, Elizabeth Smith, Chris Worth, workshop participants at the ESSLLI 2007 Workshop on New Directions in Type-Theoretic Grammar (Dublin), the Fourth

[7] For polymorphically typed operators, this is only for the lowest (ϵ) types. For $\sigma = A_0 \ldots A_n$ $(n > 0)$, the transform is $\lambda_{x_0} \ldots \lambda_{x_n}(a\ \lambda_x b(x_0) \ldots (x_n))$.

Workshop on Lambda Calculus and Formal Grammar (Nancy, 2007), the Colloque en l'honneur d'Alain Lecomte (Pauillac, 2007), the Second Workshop on Types, Logic, and Grammar (Barcelona, 2007), and colloquium audiences at the Séminaire de l'UMR 7023, CNRS/Université de Paris 8 (2008) and the Centre de Lingüística Teòrica, Universitat Autònoma de Barcelona (2008). For their help in providing the conditions that made this research possible, I am grateful to Philippe de Groote, Carlos Martin Vide, and to the Department of Linguistics and College of Humanities of Ohio State University. Some of the research reported here was supported by grant no. 2006PIV10036 from the Agència de Gestió d'Ajuts Universitaris i de Recerca of the Generalitat de Catalunya.

Appendix: The RC Schemata

H: $x : A \vdash x : A \dashv$ (x a fresh variable of type A)

A: $\vdash c : A \dashv$ (c a basic constant of type A)

M: If $\Gamma \vdash f : A \rightarrow B \dashv \Delta$ and $\Gamma' \vdash a : A \dashv \Delta'$
then $\Gamma; \Gamma' \vdash (f\ a) : B \dashv \Delta; \Delta'$

C: If $\Gamma \vdash a : A^C_B \dashv \Delta$
then $\Gamma \vdash x : A \dashv a_x : A^C_B; \Delta$ (x fresh)

R: If $\Gamma \vdash b : B \dashv a_x : A^C_B; \Delta$
then $\Gamma \vdash (a_x b) : C \dashv \Delta$

O: If $\Gamma \vdash a : A^C_B \dashv \Delta$ and $x : A, \Gamma' \vdash b : B \dashv \Delta'$
then $\Gamma; \Gamma' \vdash (a_x b) : C \dashv \Delta, \Delta'$

References

Anoun, H., Lecomte, A.: Linear grammars with labels. Formal Grammar 2006 (2007)

Bach, E., Partee, B.: Anaphora and semantic structure. In: Krieman, J., Ojeda, A. (eds.) Papers from the Parasession on Pronouns and Anaphora, pp. 1–28. Chicago Linguistic Society, Chicago (1980)

Bach, E.: Discontinuous constituents in generalized categorial grammar. NELS 11, 1–22 (1981)

Barker, C.: Continuations and the nature of quantification. Natural Language Semantics 10, 211–242 (2002)

Bernardi, R., Moortgat, M.: Continuation semantics for symmetric categorial grammar. In: Leivant, D., de Queiroz, R. (eds.), pp. 53–71 (2007)

Blackburn, P., Bos, J.: Representation and Inference for Natural Language. CSLI, Stanford (2005)

Carpenter, B.: A deductive account of scope. In: Proceeding of the Thirteenth West Coast Conference on Formal Linguistics, San Diego, CSLI, Stanford (1994)

Carpenter, B.: Type-Logical Semantics. MIT Press, Cambridge (1997)

Curien, P.-L., Herbelin, H.: The duality of computation. In: ICFP 2000, pp. 233–243 (2000)

Church, A.: A formulation of a simple theory of types. Journal of Symbolic Logic 5, 56–68 (1940)

Cooper, R.: Montague's Semantic Theory and Transformational Syntax. Ph.D. dissertation, University of Massachusetts at Amherst (1975)

Cooper, R.: Quantification and Syntactic Theory. Reidel, Dordrecht (1983)

Danvy, O., Filinski, A.: Abstracting control. In: Proceedings of the 1990 ACM Conference on Lisp and Functional Programming, pp. 151–160. ACM Press, New York (1990)

Felleisen, M.: The theory and practice of first-class prompts. In: POPL 1988, pp. 180–190. ACM Press, New York (1988)

Fischer, M.: Lambda calculus schemata. In: Proceedings of the ACM Conference Proving Assertions about Programs. SIGPLAN Notices, SIGACT News, vol. 7(1), (14), pp. 104–109 (January 1972)

Gazdar, G.: English as a context-free language. Unpublished manuscript dated April 1979, Cognitive Science Program, University of Sussex (1979)

de Groote, P.: Type raising, continuations, and classical logic. In: van Rooy, R., Stokhof, M. (eds.) Proceedings of the Thirteenth Amsterdam Colloquium. Institute for Logic, Language, and Computation, Universiteit van Amsterdam, Amsterdam, pp. 97–101 (2001)

Gallin, D.: Intensional and Higher Order Modal Logic. North-Holland, Amsterdam (1975)

Hendriks, H.: Studied Flexibility: Categories and Types in Syntax and Semantics. Ph.D. dissertation, Universiteit van Amsterdam (1993)

Henkin, L.: Completeness in the theory of types. Journal of Symbolic Logic 15, 81–91 (1950)

Keller, W.: Nested Cooper storage: the proper treatment of quantification in ordinary noun phrases. In: Reyle, U., Rohrer, C. (eds.) Natural Language Parsing and Linguistic Theories, pp. 432–447. Reidel, Dordrecht (1988)

Leivant, D., de Queiroz, R. (eds.): WoLLIC 2007. LNCS, vol. 4576. Springer, Heidelberg (2007)

Montague, R.: The proper treatment of quantification in ordinary English. In: Thomason, R. (ed.) Formal Philosophy: Selected Papers of Richard Montague, pp. 247–270. Yale University Press, New Haven (1974)

Moortgat, M.: Categorial Investigations: Logical and Linguistic Aspects of the Lambek Calculus. Ph.D. dissertation, Universiteit van Amsterdam. Foris, Dordrecht (1988)

Moortgat, M.: Generalized quantifiers and discontinuous type constructors. In: Bunt, H., van Horck, A. (eds.) Discontinuous Constituency, pp. 181–207. De Gruyter, Berlin (1996)

Moortgat, M.: Symmetries in natural language syntax and semantics. In: Leivant, D., de Queiroz, R. (eds.), pp. 264–268 (2007)

Morrill, G.: Type Logical Grammar: Categorial Logic of Signs. Kluwer, Dordrecht (1994)

Parigot, M.: $\lambda\mu$-calculus: an algorithmic interpretation of classical natural deduction. In: Voronkov, A. (ed.) LPAR 1992. LNCS, vol. 624. Springer, Heidelberg (1992)

Parigot, M.: On the computational interpretation of negation. In: Clote, P.G., Schwichtenberg, H. (eds.) CSL 2000. LNCS, vol. 1862, pp. 472–484. Springer, Heidelberg (2000)

Plotkin, G.: Call-by-name, call-by-value, and the λ-calculus. Theoretical Computer Science 1(2), 125–159 (1975)

Pollard, C.: Nonlocal dependencies via variable contexts. In: Muskens, R. (ed.) Workshop on New Directions in Type-Theoretic Grammar. ESSLLI 2007, Dublin. Revised and extended version under review for a special issue of Journal of Logic, Language, and Information (2007)

Pollard, C.: Covert movement in logical grammar. In: Pogodalla, S., Quatrini, M., Retoré, C. (eds.) Logic and Grammar: Essays Presented to Alain Lecomte on the Occasion of his 60th Birthday. LNCS/FOLLI (in press)

Shan, C.-c.: Delimited continuations in natural language: quantification and polar sensitivity. In: Continuation Workshop 2004, Venice (2004)

Stabler, E.: Remnant movement and structural complexity. In: Bouma, G., Hinrichs, E., Kruiff, G.-J., Oehrle, R. (eds.) Constraints and Resources in Natural Language, CSLI, Stanford (1999)

Vermaat, W.: Controlling Movement: Minimalism in a Deductive Perspective. Master's thesis, Utrecht University (1999)

Negative Translations and Duality: Toward a Unified Approach

Mattia Petrolo

SPHERE, Université Paris Diderot - Paris 7
mattia.petrolo@univ-paris-diderot.fr

Abstract. We address two related topics concerning recent developments of constructive classical logic. The first topic concerns the well known relationships that negative translation (a.k.a. CPS translation) establishes between classical and intuitionistic logic. We examine why they fail to give us a clear and complete picture of constructive classical logic. Secondly, we analyze some recently developed classical calculi which shed new light on negative translation and its connections with the concept of syntactical duality.

1 Introduction

1.1 Proofs-as-Programs for Classical Logic

The extension of the proofs-as-programs paradigm to classical logic sheds new light on the relationships between logic and computation. It provides a new understanding in two different ways: on one hand there are more possibilities for a logical "modelization" of computational features, on the other hand there is the discovery of new features in logic.

The fundamental discovery concerning the computational meaning of classical logic was made in [15], based on previous works by Felleisen and his collaborators. Griffin discovered that, the rules of classical logic (i.e. intuitionistic rules extended with double-negation elimination rule) can type a computational calculus equipped with control operators and that the usual negative translations from classical to intuitionistic logic correspond to what computer scientists call *continuation passing style* (CPS) translations. Further developments in this direction showed how classical logic can be used to type other well known features of imperative programming languages, namely storage operators (see [19]).

On the other hand, classical logic benefited from the input of computer science. One of the main examples of this is the discovery of polarities in logic. The first "good" constructive classical system is Girard's LC. In his article *A new constructive logic: classical logic*, Girard introduces a classical system which surprisingly keeps the good computational properties of intuitionistic logic: strong normalization and confluence. Moreover, LC is equipped with a denotational semantics. These unexpected results are obtained exploiting a polarized classical calculus and then applying the focalization algorithm.

Lecomte and S. Tronçon (Eds.): PRELUDE 2010, LNAI 6505, pp. 188–204, 2011.

1.2 Constructive Classical Systems

Since then, many different classical systems appeared, each of which pointed out a specific aspect of the constructive meaning of classical logic. If we look at these calculi from an epistemological point of view, we can distinguish two main categories. The first one includes "polarized" calculi (e.g. LC, $\lambda\mu$-calculus, LK_p^{η}...), the second one "symmetric" calculi ($\bar{\lambda}\mu\tilde{\mu}$, LK^{tq}, λ^{Sym}...). The difference between these two categories is that we achieve straightforwardly the confluence property only in the first class of calculi. Each of these two categories can be further analyzed: the polarized calculi can be divided in *explicit* and *implicit* polarized calculi. On the other hand, we can distinguish between *potential* and *actual* symmetrical calculi[1], as shown in Fig. 1.

Polarized calculi	*Explicit* (e.g. LC, LK_p^{η}, LLP ...) *Implicit* (e.g. $\lambda\mu$, LKT, LKQ ...)
Symmetric calculi	*Potential* (e.g. $\bar{\lambda}\mu\tilde{\mu}$, $\lambda\mu$-sym, LK^{tq}...) *Actual* (e.g. λ^{Sym}, ($\mathcal{T}$, $\rightarrow_{Cut}$), χ-calculus ...)

Fig. 1. A classification of constructive classical systems

So today the traditional problem about the limits of the constructivity of classical logic has become the problem to find new criteria to define the "good" computational properties for classical systems. In both approaches, it is assumed that a good computational system must be closed under a cut elimination procedure (i.e. β-reduction) and under an identity axioms expansion procedure (i.e. η-expansion). But they agree also on the fact that these requirements are generally insufficient. As a matter of fact, it is not clearly established that the confluence property is a necessary condition for "good" computational behavior in a system of classical logic.

In the present paper our aim is two folded: firstly, we want analyze how the traditional notion of negative translation can be related to the recent debate concerning the constructivization of classical logic; secondly, we shall recognize what insights negative translations can bring for the definition of what counts as a "good" system of constructive classical logic.

[1] Recent developments of constructive classical logic point out interesting links between polarized classical logic and game-semantics on one hand and symmetrical classical systems and concurrent computation on the other.

2 Negative Translations

2.1 A Brief History of Negative Translations

The first formulation of a negative translation can be found in an article that Kolmogorov published in 1925 (see [17]). His article was an explicit attempt to establish that hilbertian infinitary methods lead only to correct finitary results.

Kolmogorov presents the first (partial) axiomatization of propositional and predicate minimal logic and shows that classical propositional logic is interpretable in an intuitionistically acceptable fragment of it. In order to do so, he develops an interpretation A^K for every proposition A such that if A is classically provable, then A^K is intuitionistically provable.

The $(-)^K$-interpretation is defined by recurrence on formulas. If X is an atomic formula, then X^K is $\neg\neg X$. If A is a composed formula, $F(\phi_1, \phi_2, ..., \phi_k)$, then A^K is $\neg\neg F(\phi_1^K, \phi_2^K, ..., \phi_k^K)$. Kolmogorov considers explicitly only the $\neg$ and $\rightarrow$ connectives, but his interpretation can be extended straightforwardly to the connectives $\vee$ and $\wedge$. It is clear that this interpretation is far from being minimal (e.g. for negation the definition gives $(\neg A^K) = \neg\neg\neg A^K$, when we know that $(\neg A^K) = \neg A^K$ suffices), but it is extremely systematic. Kolmogorov also made explicit suggestions concerning the treatment of predicate logic, even though his treatment of quantifiers was incomplete.

Few years after the publication of Kolmogorov's paper, Glivenko presents an argument in order to disarm the criticism of Barzin and Errera concerning the fact that intuitionistic logic was a three valued logic and that it was inconsistent (see [13]). Glivenko proves that

1. *If A is provable in classical propositional logic, then $\neg\neg A$ is provable in intuitionistic propositional logic;*
2. *If $\neg A$ is provable in classical propositional logic, then it is also provable in intuitionistic propositional logic.*

So, as pointed out in [2], this is not exactly a negative translation (as Kolmogorov's is), because it does not translate subformulas of A. But it is strong enough to show that the classical and intuitionistic propositional systems are equiconsistent.

Glivenko's result cannot be extended to quantified formulas. Such an extension is given by Kuroda (see [20]). In fact, his strategy is exactly to construct A^K from A by putting $\neg\neg$ before the whole formula and after every universal quantifier. But, in order to obtain the Kuroda negative translation, it is required to use intuitionistic logic (not only minimal logic).

It is worth noting that Kolmogorov and Glivenko's systems are not equivalent, notably the former rejects *ex falso quodlibet*, while the latter accepts it.

Both works of the Russian mathematicians failed to give a complete and correct treatment of propositional logic. The first published work to give a full account of negative translation for first order logic was [14]. Gödel showed also a partial improvement in his interpretation, by adding the double-negation only on atoms, $\rightarrow$, $\vee$ and $\exists$. Gentzen found independently from Gödel an equivalent translation, withdrawn upon learning of Gödel's paper and finally published only in 1936 (see [11]). In fact, Gentzen's interpretation is a slightly optimized version of Gödel's one. It shows how it is possible to add a double-negation only on atoms, $\vee$ and $\exists$, the "crucial" intuitionistic connectives. We present what is usually called the Gödel-Gentzen negative translation:

$$\begin{array}{rl}
\text{If } X \text{ is atomic, then } X^G & \text{is } \neg\neg X \\
(A \wedge B)^G & \text{is } A^G \wedge B^G \\
(A \vee B)^G & \text{is } \neg(\neg A^G \wedge \neg B^G) \\
(A \rightarrow B)^G & \text{is } A^G \rightarrow B^G \\
(\neg A)^G & \text{is } \neg A^G \\
(\forall x A)^G & \text{is } \forall x A^G \\
(\exists x A)^G & \text{is } \neg \forall x \neg A^G
\end{array}$$

Gödel extends his negative translation to Number Theory; such an extension shows that Peano Arithmetic (PA) is translatable into Heyting Arithmetic (HA). This completes and generalizes Kolmogorov's result from 1925, which at the time was known to neither Gentzen nor Gödel. By carrying out such a translation, Gödel gives the first relative consistency proof of classical number theory with respect to intuitionistic one.

A first step toward a proof-theoretical understanding of negative translations was carried out through Friedman's A-translation (see [10]). Such a translation explains how to recover the proof of the original sentence once we have its double-negated version. Moreover, a combination of negative translation with the A-translation provides a simpler proof of a well-know conservative result due to Kreisel. Let A and B are intuitionistic formulas, where no free variable of B is quantified in A. The translation $(-)^A$ is defined by replacing each atomic subformula X of A by $X \vee A$. I.e.

$$\begin{array}{rcl}
X^A & = & X \vee A \\
(A \wedge B)^A & = & A^A \wedge B^A \\
(A \vee B)^A & = & A^A \vee B^A \\
(A \rightarrow B)^A & = & A^A \rightarrow B^A \\
(\forall x A)^A & = & \forall x A^A \\
(\exists x A)^A & = & \exists x A^A
\end{array}$$

Friedman considered $\perp$ an atomic formula as well, hence it is replaced with $\perp \vee A$ (which is equivalent to A). Note that $\neg B$ is defined as an abbreviation for $B \rightarrow \perp$, hence $(\neg B)^A = B^A \rightarrow A$. A-translation can be combined both with Gödel-Gentzen and Kolmogorov negative translation. By doing so, it is possible to generalize Gödel's result for number theory and prove Π_2^0-conservativity of many classical theories T over their intuitionistic counterpart T_i. As a corollary of this, one gets that T has the same provably recursive functions as T_i.

In particular, Friedman shows that if PA $\vdash \forall x \exists y f(x, y) = 0$, then HA $\vdash \forall x \exists y f(x, y) = 0$. This result is today known as the Kreisel-Friedman theorem. It is worth noting that such a result not only pointed out the possibility to extract a "constructive" meaning from proofs based on classical logic, but also provides the concrete means to carry out such an extraction. It is actually an evidence of the constructivity of classical logic long before the Curry-Howard isomorphism was extended to it.

2.2 CPS and the Computational Meaning of Negative Translations

The idea of continuations and continuation-passing-style translation originated with work in the denotational semantics of programming languages[2]. A *continuation* in denotational semantics is a value that needs to be applied to a value and/or store, and returns a final answer, the result of the entire computation. Programs fragments are viewed as continuation transformers, which take continuations as arguments and return other continuations. Within this framework, one can define the mathematical semantics of nonfunctional languages.

The next step toward the understanding of continuations has been the development of specific operators allowing to access the current continuation. The (nonfunctional) control operator $\mathcal{C}$ was invented explicitly to achieve this effect. Its evaluation semantics can be expressed through another nonfunctional control operator , $\mathcal{A}$, and two evaluation rules:

$$\begin{array}{rcl} E[\mathcal{C}(M)] & \longrightarrow & M(\lambda x.\mathcal{A}(E[x])) \\ E[\mathcal{A}(M)] & \longrightarrow & M \end{array}$$

First rule explains how the continuation E gets captured. $\mathcal{C}$, when applied to a term M, packages up its current evaluation context in a function, and resumes evaluating M applied to that function. The effect is to allow M to do what it wants when it is finished and wants to exit - it can resume execution at the original evaluation context, or resume at another evaluation context if it has others available, etc. $\mathcal{A}$ simply discards away its current evaluation context and proceeds with evaluating its argument in the empty context. In the terminology of abstract machines, the continuation E can be conceived as the stack of all things that are waiting to be done in the future. Mathematically, a continuation can be seen as a functional applying its function argument to a stack of arguments. It turns out that many nonfunctional control operations can be expressed with these simple additions to a functional language.

Fischer introduced in [9] the continuation-passing-styles (CPS) translations by exploiting the basic features of continuation semantics; such a translation, when applied to a program, converted that program to one in which the *evaluation order* was specified not externally, but explicitly in the program.

In [27], Plotkin extended this work by showing that one could, in a certain sense, simulate a call-by-value programming language in a call-by-name language, and vice-versa. More precisely, given a language $\mathcal{L}$ (the object-language) and a language $\mathcal{M}$ (the meta-language), Plotkin gives a translation $\bullet : \mathcal{L} \longrightarrow \mathcal{M}$ such that for any term X of $\mathcal{L}$, if $X \rightarrow b$, then $X^{\bullet} \rightarrow b'$, where b' is a value which mimics in a precise sense the operational semantics of b. Plotkin's translation can be defined as follows:

[2] Our presentation is based on [24] where classical CPS translations are analyzed in depth for the first time.

$$
\begin{aligned}
x^{\bullet} &= \lambda k(x)k \\
(\lambda x u)^{\bullet} &= \lambda k(k)\lambda x u^{\bullet} \\
((u)v)^{\bullet} &= \lambda k(u^{\bullet})\lambda m((m)v^{\bullet})k
\end{aligned}
$$

By extending Plotkin's translation to types, it becomes clear that it is a version of Friedman's A-translation[3]:

$$
\begin{aligned}
X^{\bullet} &= X \\
(A \to B)^{\bullet} &= \neg\neg A^{\bullet} \to \neg\neg B^{\bullet}
\end{aligned}
$$

Thus, by exploiting a CPS-translation, Plotkin showed that one could simulate a call-by-name (CBN) lambda-calculus in a call-by-value (CBV) lambda-calculus and that the converse is also true. Plotkin refers to his translation from $\mathcal{L}$ to $\mathcal{M}$ as providing a definitional interpreter for a programming language $\mathcal{L}$ in a programming language $\mathcal{L}$. That is, such a translation provides a way of understanding programs in $\mathcal{L}$ without understanding the evaluation order on programs in $\mathcal{L}$. [8] extended the CPS-translation idea to encompass the operator *control* ($\mathcal{C}$), showing that one could translate a functional language with instances of $\mathcal{C}$ into a language without such instances, such that the original program and the translated program give equal values.

Griffin observed that the typing constraints inferred by reduction are met exactly when M and $\mathcal{C}(M)$ in the rules above have type $(A \to R) \to R$ (for any type R) and A, respectively, i.e.

$$
\frac{\Gamma \vdash M : (A \to R) \to R}{\Gamma \vdash \mathcal{C}(M) : A}
$$

If R is interpreted as $\bot$ and $\neg A$ is defined as $A \to \bot$, we obtain the classical rule of $\neg\neg$-elimination

$$
\frac{\Gamma \vdash M : \neg\neg A}{\Gamma \vdash \mathcal{C}(M) : A}
$$

Griffin then observed that call-by-value CPS-translations translated a program typed in typed λ-calculus $+\ \mathcal{C}$ to the typed λ-calculus, in such a way that the original program computed to an integer b if and only if the translated program also computed to the same integer. Moreover, the translation upon types induced by such a translation corresponds to a double-negation translation. In order to achieve this result, Griffin used the implicational fragment of the A-translation combined with a modified Kuroda double-negation translation. Thus, exploiting the Curry-Howard isomorphism, Griffin discovered that to change negative translation corresponds to change the evaluation order. >From a semantical point of view, Griffin proved that CPS-translation of programs with $\mathcal{C}$ preserves their

[3] As showed in [23] the "dual" of such a translation in terms of linear analysis is given by Krivine's translation (see [18]), a slight amelioration of Gödel-Gentzen translation. It is defined as follows: $X^* = \neg X \mid (A \to B)^* = A^* \to B^*$. So, the formula A^* is obtained by putting only one $\neg$ before every atomic subformula of A.

semantics (i.e. the program in the object-language is semantically equivalent to the program in the meta-language).

All the above-mentioned CPS translations (and continuation semantics) comprise a notion of value even for the call by name variants. An effect of such a choice is that they do not validate the η-reduction. In order to overcame such a difficulty, Lafont, Streicher and Reus define a translation that does not admit a basic notion of value but, instead, a basic notion of continuation (see [21]). In a type-theoretic setting, Lafont-Streicher-Reus translation can be defined as follows:

$$\begin{aligned} X^L &= \neg X \\ (A \wedge B)^L &= A^L \vee B^L \\ (A \vee B)^L &= (A^L \wedge B^L) \\ (A \rightarrow B)^L &= \neg A^L \wedge B^L \\ (\neg A)^L &= \neg A^L \end{aligned}$$

Note that such a translation is different form Gödel-Gentzen and Kolmogorov negative translations which correspond to a call-by-name CPS translation with values and a call-by-value one, respectively[4]. From a categorical point of view, the continuation semantics of Lafont-Streicher-Reus negative translation gives rise to a cartesian closed category called the category of "negated domains". Such a development was one of the main inspiration for the discovery of control categories (see [28]), a well-behaved categorical semantics for languages with continuation-like control construct, such as Felleisen's $\mathcal{C}$ or Parigot $\lambda\mu$[5].

Nonetheless, as remarked by Danvy *et al.* in [6], "Yet no standard CPS-transformation algorithm has emerged, and this missing piece contributes to maintaining continuations, CPS, and CPS transformations as mystifying artifacts [...] in the land of programming and programming languages".

2.3 Negative Translations and Constructive Classical Logic

By now, it should be clear that negative translations bring an understanding of classical logic through intuitionistic logic. The extension of Curry-Howard isomorphism to classical logic enables us to reconsider such a position.

As showed in [29], in classical non-confluent *symmetric* systems there exist some normal forms that are not reachable in strongly normalizing and confluent *polarized* systems. What is more, [3] gives an example of a classical proof meaningful from a computational viewpoint but which cannot be extracted by a

---

[4] It is possible to add to ^{L}-translation $\forall$, but not $\exists$: in a sequent calculus setting the $\exists$ left rule cannot be correctly translated.

[5] The $\lambda\mu$-calculus is the first clear proof-theoretical account of the computational meaning of classical logic (see [26]). In the computational framework we are considering here, Parigot's calculus can be regarded as a prototypical call-by-name language with control primitives for handling continuations. In this respect, it is similar to programming languages with Felleisen's $\mathcal{C}$ operator, except that the latter language is call-by-value.

double-negation translation. Indeed, in these non-confluent calculi it is possible to choose how to reduce cuts *during* the cut-elimination process. This shows that a notion of non-determinism present in classical logic is not captured by negative translations.

These facts seem to suggest the need for a re-examination of the traditional notion of negative translation. A first obvious question is: which restrictions are enforced by negative translations so to lead to a single normal form from a classical proof? Then, if we turn our attention to the *dynamical* aspects of classical logic, we should try to understand what are the relationships between negative translations and the process of cut-elimination. In particular, do negative translations correspond to particular strategies of how to eliminate cuts? Let us give a concrete example based on [30] in order to address the answers to these tricky questions.

In a sequent calculus setting, the Gödel-Gentzen negative translation given in paragraph 2.1 is a translation of a classical proof having an end-sequent

$$\begin{matrix}\vdots\\ \Gamma \vdash \Delta\end{matrix}$$

to an intuitionistic proof with end-sequent

$$\begin{matrix}\vdots\\ \Gamma^G, \neg\Delta^G \vdash\end{matrix}$$

preserving the "structure" of the classical proof.

Let us now translate the $\wedge_R$ rule through the $(-)^G$ translation. When translating from a classical proof

$$\dfrac{\begin{matrix}\vdots\\ \vdash B\end{matrix} \qquad \begin{matrix}\vdots\\ \vdash C\end{matrix}}{\vdash B \wedge C}\,\wedge_R$$

we have by induction hypothesis two intuitionistic proofs ending in

$$\begin{matrix}\vdots\\ \vdash B\end{matrix} \qquad \begin{matrix}\vdots\\ \vdash C\end{matrix}$$

So we can construct the following translated proof for $\wedge_R$:

$$\dfrac{\dfrac{\dfrac{\dfrac{\begin{matrix}\vdots\\ \neg B^G \vdash\end{matrix}}{\vdash \neg\neg B^G}\,\neg_R \qquad \begin{matrix}\vdots\\ \neg\neg B^G \vdash B^G\end{matrix}}{\vdash B^G}\,Cut \qquad \dfrac{\dfrac{\begin{matrix}\vdots\\ \neg C^G \vdash\end{matrix}}{\vdash \neg\neg C^G}\,\neg_R \qquad \begin{matrix}\vdots\\ \neg\neg C^G \vdash C^G\end{matrix}}{\vdash C^G}\,Cut}{\vdash B^G \wedge C^G}\,\wedge_R}{\neg(B^G \wedge C^G) \vdash}\,\neg_L$$

In order to obtain the translated proof, the property that the sequent $\neg\neg(-)^G \vdash (-)^G$ is always derivable in LJ is needed. Remark that in the example, the $(-)^G$ translation introduces new cuts not present in the classical version. Following [30], we replace a cut introduced by the $(-)^G$ translation with a shorter rule $(\neg\neg_R)$, i.e.

$$\dfrac{\Gamma \vdash \neg\neg B \qquad \neg\neg B \vdash B}{\Gamma \vdash B}\,Cut \qquad \rightsquigarrow \qquad \dfrac{\vdash \neg\neg B}{\vdash B}\,\neg\neg_R$$

When trying to translate the instances of the cut-rule we stumble upon a problem: it can be $(-)^G$ translated in two different ways. The following cut-rule

$$\dfrac{\begin{matrix}\pi_1\\ \vdots\\ \vdash B\end{matrix} \qquad \begin{matrix}\pi_2\\ \vdots\\ B \vdash\end{matrix}}{\vdash}\,Cut$$

can be translated either as

$$\dfrac{\begin{matrix}\pi_1^G\\ \vdots\\ \neg B^G \vdash\end{matrix} \qquad \dfrac{\begin{matrix}\pi_2^G\\ \vdots\\ B^G \vdash\end{matrix}}{\vdash \neg B^G}\,\neg_R}{\vdash}\,Cut \qquad \text{or} \qquad \dfrac{\dfrac{\dfrac{\begin{matrix}\pi_1^G\\ \vdots\\ \neg B^G \vdash\end{matrix}}{\vdash \neg\neg B^G}\,\neg_R}{\vdash B^G}\,\neg\neg_R \qquad \begin{matrix}\pi_2^G\\ \vdots\\ B^G \vdash\end{matrix}}{\vdash}\,Cut$$

i.e. as a *left* or a *right*-translation of a cut, respectively. Let us now turn to the problem to simulate all classical cut-reduction sequences through the $(-)^G$ translation.

One of the more comprehensive studies of confluent variants of classical systems is carried out in [7] through the system LK^{tq} (i.e. the "colored" version of LK), which is strongly normalizing and confluent. The tq-protocol establishes two different deterministic reduction-strategies, depending on the chosen color for a formula. Once the color of a formula has been chosen, the normal form is uniquely determined. Such a choice is not equivalent to choose a strategy for cut-elimination, but it rather corresponds to the way how cuts are reduced. A crucial technical feature of LK^{tq} is that once a "color" (*right* or *left*) for a formula A has been chosen, all occurrences of A in the proof must have the same color. By exploiting the system LK^{tq}, we are able to analyze the behavior of negative translations during the cut-elimination process. For example, the logical cut

$$\dfrac{\dfrac{\begin{matrix}\pi_1\\ \vdots\\ \vdash B\end{matrix}}{\vdash B \vee C}\,\vee_{R_1} \qquad \dfrac{\begin{matrix}\pi_2\\ \vdots\\ B \vdash\end{matrix} \qquad \begin{matrix}\pi_3\\ \vdots\\ C \vdash\end{matrix}}{B \vee C \vdash}\,\vee_L}{\vdash}\,Cut$$

can be reduced to

$$\dfrac{\begin{matrix}\pi_1\\ \vdots\\ \vdash B\end{matrix} \qquad \begin{matrix}\pi_2\\ \vdots\\ B \vdash\end{matrix}}{\vdash}\,{\scriptstyle Cut}$$

and, assuming that B is not freshly introduced in π_1, the cut-rule can be eliminated. In LK^{tq} this behavior corresponds to the '$\leftarrow$' color annotation for B, i.e.

$$\dfrac{\dfrac{\begin{matrix}\pi_1\\ \vdots\\ \vdash \overleftarrow{B}\end{matrix}}{\vdash \overleftarrow{B} \vee C}\,{\scriptstyle \vee_{R_1}} \qquad \dfrac{\begin{matrix}\pi_2\\ \vdots\\ \overleftarrow{B} \vdash\end{matrix} \quad \begin{matrix}\pi_3\\ \vdots\\ C \vdash\end{matrix}}{\overleftarrow{B} \vee C \vdash}\,{\scriptstyle \vee_L}}{\vdash}\,{\scriptstyle Cut}$$

As showed in [30], the behavior of this proof can be simulated by the $(-)^G$ translation as follows:

$$\dfrac{\dfrac{\dfrac{\dfrac{\begin{matrix}\pi_1^G\\ \vdots\\ \neg B^G \vdash\end{matrix}}{\neg B^G \wedge \neg C^G \vdash}\,{\scriptstyle \wedge_{L_1}}}{\vdash \neg(\neg B^G \wedge \neg C^G)}\,{\scriptstyle \neg_R}}{\neg\neg(\neg B^G \wedge \neg C^G) \vdash}\,{\scriptstyle \neg_L} \qquad \dfrac{\dfrac{\dfrac{\dfrac{\begin{matrix}\pi_2^G\\ \vdots\\ B^G \vdash\end{matrix}}{\vdash \neg B^G}\,{\scriptstyle \neg_R} \quad \dfrac{\begin{matrix}\pi_3^G\\ \vdots\\ C^G \vdash\end{matrix}}{\vdash \neg C^G}\,{\scriptstyle \neg_R}}{\vdash \neg B^G \wedge \neg C^G}\,{\scriptstyle \wedge_R}}{\neg(\neg B^G \wedge \neg C^G) \vdash}\,{\scriptstyle \neg_L}}{\vdash \neg\neg(\neg B^G \wedge \neg C^G)}\,{\scriptstyle \neg_R}}{\vdash}\,{\scriptstyle Cut}$$

This proof reduces in three steps to

$$\dfrac{\begin{matrix}\pi_1^G\\ \vdots\\ \neg B^G \vdash\end{matrix} \qquad \dfrac{\begin{matrix}\pi_2^G\\ \vdots\\ B^G \vdash\end{matrix}}{\vdash \neg B^G}\,{\scriptstyle \neg_R}}{\vdash}\,{\scriptstyle Cut}$$

In this proof $\neg B^G$ is freshly introduced so the cut-rule can only be reduced to the left, exactly as the color-annotation prescribes.

Let us now simulate the opposite color-annotation for B, i.e.

$$\dfrac{\dfrac{\begin{matrix}\pi_1\\ \vdots\\ \vdash \overrightarrow{B}\end{matrix}}{\vdash \overrightarrow{B} \vee C}\,{\scriptstyle \vee_{R_1}} \qquad \dfrac{\begin{matrix}\pi_2\\ \vdots\\ \overrightarrow{B} \vdash\end{matrix} \quad \begin{matrix}\pi_3\\ \vdots\\ C \vdash\end{matrix}}{\overrightarrow{B} \vee C \vdash}\,{\scriptstyle \vee_L}}{\vdash}\,{\scriptstyle Cut}$$

It turns out that such a color-annotation cannot be simulated through the $(-)^G$ translation: it would not respect the constraints given by colored formulas. Urban and Ratiu show that the behavior of the new color annotation can be simulated through the $(-)^K$ translation we have defined in paragraph 2.1. Once we have applied the $(-)^K$ translation, we obtain the following proof after four steps of cut reduction:

$$\dfrac{\dfrac{\dfrac{\begin{matrix}\pi_1^K\\ \vdots\\ \neg B^K \vdash\end{matrix}}{\vdash \neg\neg B^K}\,\neg R}{\vdash B^K}\,\neg\neg R \qquad \begin{matrix}\pi_2^K\\ \vdots\\ B^K \vdash\end{matrix}}{\vdash}\,Cut$$

It is clear that such a proof cannot be reduced on the left, because the cut-formula is freshly introduced. So we must reduce the cut to the right, exactly as the color annotation prescribed.

This example provides an evidence to support the claim that we need more than a single negative translation in order to capture the whole computational meaning of classical proofs.

2.4 Limits of Standard Negative Translations

Negative translations aim to give a finer meaning to classical logic through intuitionistic logic. From this viewpoint, intuitionistic logic tries to "deconstruct" [6] through negative translations standard classical operations in order to look for their actual constructive meaning.

Nonetheless, when we consider intuitionistic logic as a tool of deconstructive analysis for classical logic, some of its limits appear clearly. First of all, there are cases in which a single negative translation is ambiguous. The example we have seen in paragraph 2.3 shows that the $(-)^G$ translation cannot correctly simulate the whole cut-elimination process for the classical system LK^{tq}. Another limit of an intuitionistic analysis of classical logic appears if we focus our attention on the sequent calculus. As remarked in [7], intuitionistic sequent calculus is based on an *ad hoc* dissymetrization of LK, not better than the dual dissymetrization, as we shall see in next section. Finally, from a technical point of view, negative translations have several flaws pointed out by Girard in [12]: 1) they are not compatible with substitution; 2) negation is not involutive; 3) disjunction is not associative.

On the one hand, these limits lead to think that intuitionistic logic is not the right framework in which the constructive meaning of classical logic should be analyzed. On the other hand, they suggest that we are not completely on the wrong track and that we need some (though crucial) refinement to have a deeper and unified view of constructive classical logic. Such a refinement is

[6] The expression is borrowed from [7].

obtained through linear logic. Sometimes it is claimed that linear logic (LL) represents a refined analysis of intuitionistic logic. More precisely, Girard discovered a decomposition of the usual intuitionistic operations: for example, the intuitionistic implication $A \to B$ becomes $(!A) \multimap B$ in LL. The fundamental difference between linear and intuitionistic negation is that the former is involutive, the latter is not.

Thus, one of the main reasons for which LL can be seen as a good candidate to achieve a correct deconstructive analysis for classical logic, is precisely its deep investigation and accurate treatment of negation. Linear logic allows a refinement of the connective $\neg$ as used in negative translations. In particular, it allows us to separate the "real" operation of negation from a simple change of side of the symbol $\vdash$ in a sequent.

3 Polarization and Duality

3.1 Co-intuitionistic Logic and Classical Logic

As already remarked, the intuitionistic system LJ impose an *ad hoc* dyssimetrization of the classical calculus LK. It is clearly possible to adopt a dual formulation of LJ. At the level of sequents it means to pass from $\Gamma \vdash A$ to $A \vdash \Gamma$, where A is at most one formula. Let us call co-LJ the system obtained by dualizing the intuitionistic sequent calculus LJ. Such a calculus was introduced in [5].

Let us analyze the formal relationships between LJ and co-LJ. A translation from LJ to co-LJ is easily defined in the following way:

$$\begin{aligned} \overline{X} &= X \\ \overline{A \wedge B} &= \overline{A} \vee \overline{B} \\ \overline{\neg A} &= \neg \overline{A} \\ \overline{A \vee B} &= \overline{A} \wedge \overline{B} \end{aligned}$$

From this translation we obtain the following property:

Duality between LJ and co-LJ. $\Gamma \vdash_{LJ} A$ *if and only if* $\overline{A} \vdash_{co-LJ} \overline{\Gamma}$.

Remark that such a translation gives an involution on formulas: $\overline{\overline{A}} = A$. As proved in [5] there is no loss of provability by moving from LK to co-LJ[7].

In order to show some interesting proof-theoretical relationships between LJ/co-LJ and LK, we need again the technical apparatus developed in [7] where classical logic is "constructivized" through linear logic. Remember that LK^{tq} can be divided in two dual subsystems both complete with respect to classical provability:

$$\Gamma \vdash_{LKQ} \Delta; A \iff \overline{A}; \Delta \vdash_{LKT} \overline{\Gamma}$$

[7] As showed in [22], LJ and co-LJ have an identical, computational behavior. Exploiting this fact, Laurent obtains a proof-net syntax for both calculi.

where $;A$ is called the *stoup* and contains at most one formula. Then, the relationships between intuitionistic and classical systems can be spelled out as follows:

Translations

If $\Gamma \vdash_{LKQ} \Delta; A$ then $\Gamma, \neg\Delta \vdash_{LJ} A$

If $A; \Gamma \vdash_{LKT} \Delta$ then $A \vdash_{co-LJ} \neg\Gamma, \Delta$

Remark that a flaw of standard negative translations becomes finally evident through the linear analysis of classical logic carried out in [7]. LJ and co-LJ dissymetrizations make a "connective-preserving" translation of LK impossible: LJ lacks a *multiplicative* $\vee$ and co-LJ lacks a *multiplicative* $\wedge$.

3.2 Polarization and Negative Translations

Polarization was originally discovered in a linear logic setting (see [1]), but here we shall give a polarized version of intuitionistic, co-intuitionistic and classical logic based on [31], [22].

Intuitionistic Polarization. Positive formulas P, Q are obtained through positive atoms X, Y combined with positive connectives

$$P, Q ::= X \mid P \otimes Q \mid P \oplus Q \mid \neg_{+} P$$

In terms of provability, there is a simple relationship between positive formulas and ordinary intuitionistic logic. Let $\mid P \mid$ be the operator that converts a positive formula to an ordinary formula:

$$\begin{array}{rcl} \mid X \mid & = & X \\ \mid \neg_{+} P \mid & = & \neg \mid P \mid \\ \mid P \otimes Q \mid & = & \mid P \mid \wedge \mid Q \mid \\ \mid P \oplus Q \mid & = & \mid P \mid \vee \mid Q \mid \end{array}$$

Co-intuitionistic Polarization. Negative formulas N, M are obtained through negative atoms $\overline{X}, \overline{Y}$ combined with negative connectives:

$$N, M ::= \overline{X} \mid N \mathbin{\&} M \mid N \parr M \mid \neg_{-} N$$

Such a polarization correspond to co-intuitionistic logic.

In order to be able to let intuitionistic and co-intuitionistic logic interact, we add a pair of mediating connectives called *shifts*:

$$\begin{array}{l} P, Q ::= \cdots \mid {\downarrow} N \\ N, M ::= \cdots \mid {\uparrow} P \end{array}$$

We obtain *Propositional Polarized Logic* (PPL, see [31]). Classical logic is easily obtained from PPL, by defining an operator $\mid - \mid$ that "forgets" polarities, converting a polarized formula to an ordinary propositional formula.

$$\begin{array}{rcccl} & & \mid X \mid & = & X \\ & & \mid \overline{X} \mid & = & X \\ \mid A \otimes B \mid & = & \mid A \,\&\, B \mid & = & \mid A \mid \wedge \mid B \mid \\ \mid A \oplus B \mid & = & \mid A \parr B \mid & = & \mid A \mid \vee \mid B \mid \\ & & \mid \neg_{+} P \mid & = & \neg \mid P \mid \\ & & \mid \neg_{-} N \mid & = & \neg \mid N \mid \\ & & \mid \downarrow N \mid & = & \neg \mid N \mid \\ & & \mid \uparrow P \mid & = & \neg \mid P \mid \end{array}$$

Following this characterization each formula has infinitely many polarizations.

Zeilberger proves that PPL and classical logic are interchangeable in terms of provability. The difference is that polarization impose a particular strategy for proving a classical formula. Exploiting the focalization algorithm (see [1]), Zeilberger proves that all strategies are acceptable. Thus, by defining different polarizations of the classical connectives, it is possible to reconstruct different negative translations. For example, by polarizing a formula into the purely positive fragment of PPL, gives us Glivenko's theorem; by polarizing a formula into the purely negative fragment gives us the Lafont, Reus, and Streicher negative translation.

The Gödel-Gentzen translation can be obtained through the following polarization

$$\begin{aligned} X^* &= \downarrow\uparrow X^+ \\ (A \wedge B)^* &= A^* \otimes B^* \\ (A \vee B)^* &= \downarrow(\uparrow A^* \parr \uparrow B^*) \\ (\neg A)^* &= \neg_{+} A^* \end{aligned}$$

Thus, the focusing completeness theorem establish that each of the infinitely many polarizations results in a negative translation.

3.3 Advantages of the Polarized Approach

Let us spell out what we have gained through a polarized analysis of negative translations.

Firstly, from a logical viewpoint, advantages of an approach based on polarizations are clear. Such a translation is compatible with substitution, it has an involutive negation and finally enjoys associativity of $\wedge$ and $\vee$. Nonetheless, from a technical point of view, it should be noted that finding a meaningful equivalence relation between syntactically different negative translations is still an open problem.

Secondly, from a computational point of view, polarization allows us to make a bridge between CPS translations and evaluation order in programming. In fact, what we obtain by applying the Curry-Howard isomorphism to PPL is a computational system where the evaluation order is reflected at the level of types. Then, a further analysis of classical polarized and focalized systems leads to relate positive and negative polarities to eager and lazy computational behaviors.

Finally, from an epistemological point of view[8], we have reached the unifying view that we were looking for: we have a simple characterization of all negative translations in terms of polarities. What is more, polarization give us a more clever idea of what is the computational meaning of classical logic, by showing that there is not a single constructive content of a classical theorem, because there are many different ways to polarize a classical proposition.

To summarize, if linear logic is thought as intuitionistic logic *with* ("classical") dualities[9], we can say that polarized logic is classical logic *with* negative translations. Such negative translations are now completely internalized, i.e. defined on fragments of the same unified logic. As remarked in paragraph 2.2, CPS translations, the computational counter-parts of negative translations, are also defined in terms of internal operations. We can conclude that polarized logic is actually the correct logic in which it is possible to formalize CPS translations.

4 Conclusion

The traditional view that classical logic is symmetrical was pointed out for the first time by Gentzen : "There exists complete symmetry between $\wedge$ and $\vee$, $\forall$ and $\exists$. All of the connectives have, to a large extent, equal status in the system; no connective ranks notably above any other connective"[10].

More recently, classical symmetries also appeared in computer science, for example between a program and its evaluation context (see [4]).

Nonetheless, in the de-constructive framework we have described in the last section of the present paper, such a view can be challenged: classical symmetries have disappeared, at least partially. It is usually claimed that there exists a symmetry in classical sequent calculus between the structural treatment of formulas in the hypothesis and formulas in the conclusion. However, when we analyze the dynamics of classical systems, it should be noted that such a symmetry is effective only for atomic formulas. In fact, during cut-elimination process, the reversibility properties for compound formulas break this symmetry. Polarities provide us a "natural" way to break these symmetries and define strongly normalizing and confluent classical systems.

Accepting the dissymetrization of classical logic has deep consequences when a proof is considered as the place where two actions of opposite polarity *interact*. Such a departure from the traditional view of logic is made possible exploiting the linear analysis of classical logic. In the linear deconstructive setting, nega-

[8] Zeilberger's work opens another interesting philosophical line of research concerning the meaning of logical constants. The evaluation and the study of the consequences of such an analysis go beyond the scope of the present paper and shall be discussed elsewhere.

[9] Such a view is advocated for example in [25] or in [23].

[10] Cf. [11], p. 259.

tion becomes a relation defined *modulo* De Morgan's laws. Philosophically, as remarked by Joinet in [16], this theoretical change is a decisive one: it corresponds to shift from a view of negation as a trivial connective to a conception of *negation as an act*. From this perspective, for example, the law of excluded middle $A \vee \neg A$ is far from being symmetrical: when analyzing the computational behavior of proofs, one formula *act*, while its negation *react*. In this way it is possible to define two dual alternate phases of interaction.

A natural extension of this point of view should lead to both philosophical and technical improvements. In particular, this treatment of classical logic opens the door to a new *game-theoretical* interpretation of negative translations based on the cut-elimination process. What is more, it could bridge different approaches to logic, among which is Ludics.

References

[1] Andreoli, J.-M.: Logic programming with focusing proofs in linear logic. Journal of Logic and Computation 2(3), 297–347 (1992)

[2] van Atten, M.: The Development of Intuitionistic Logic, The Stanford Encyclopedia of Philosophy (Summer 2009 Edition). Zalta, E.N. (ed.) (2009), ¡`http://plato.stanford.edu/archives/sum2009/entries/intuitionistic-logic-development`¿

[3] Barbanera, F., Berardi, S., Schivalocchi, M.: "Classical" Programming-with-Proofs in λ-sym: An Analysis of Non-Confluence. In: Ito, T., Jayaraman, K. (eds.) TACS 1997. LNCS, vol. 1281, pp. 365–390. Springer, Heidelberg (1997)

[4] Curien, P.-L., Herbelin, H.: The Duality of Computation. In: Proc. Int. Conf. on Functional Programming. ACM press, New York (2000)

[5] Czermak, J.: A remark on Gentzen's calculus of sequents. Notre Dame Journal of Formal Logic 18, 471–474 (1977)

[6] Danvy, O., Millikin, K., Nielsen, L.R.: On one-pass CPS transformations. J. Funct. Program. 17(6), 793–812 (2007)

[7] Danos, V., Joinet, J.-B., e Schellinx, H.: A New Deconstructive Logic: Linear Logic. Journal of Symbolic Logic 62(3), 755–807 (1997)

[8] Felleisen, M., Friedman, D., Kohlbecker, E., Duba, B.: Reasoning with Continuations. In: LICS 1986, pp. 131–141 (1986)

[9] Fischer, M.J.: Lambda calculus schemata. In: Proceedings of an ACM Conference on Proving Assertions about Programs, pp. 104–109 (1972)

[10] Friedman, H.: Classically and Intuitionistically Provable Recursive Functions. In: Bird, R.S., Woodcock, J.C.P., Morgan, C.C. (eds.) MPC 1992. LNCS, vol. 669, pp. 21–27. Springer, Heidelberg (1993)

[11] Gentzen, G.: The Collected Papers of Gerhard Gentzen. In: Studies in Logic and the Foundations of Mathematics, North-Holland, Amsterdam (1969)

[12] Girard, J.-Y.: A New Constructive Logic: Classical Logic. Mathematical Structures in Computer Science 1(3), 255–296 (1991)

[13] Glivenko, V.: Sur quelques points de la logique de M. Brouwer. Acadmie Royale de Belgique, Bulletin de la classe des sciences 5(15), 183–188 (1929)

[14] Gödel, K.: Zur intuitionistischen Arithmetik und Zahlentheorie. Ergebnisse eines mathematischen Kolloquiums 4, 286–295 (1933)

[15] Griffin, T.: A Formulae-as-Types Notion of Control. In: Principles of Programming Languages, pp. 47–58. ACM Press, New York (1990)

[16] Joinet, J.-B.: Sur la ngation, Manuscript communicated by the author in winter 2009, 20 pages (2009)
[17] Kolmogorov, A.: O principe tertium non datur. Matematiceskij Sbornik 32, 646–667 (1925)
[18] Krivine, J.-L.: Oprateurs de mise en mmoire et traduction de Gödel. Arch. Math. Logic 30(4), 241–267 (1990)
[19] Krivine, J.L.: Classical logic, storage operators and second-order lambda-calculus. Ann. Pure Appl. Logic 68(1), 53–78 (1994)
[20] Kuroda, S.: Intuitionistische Untersuchungen der formalistischen Logik. Nagoya Math. 3, 35–47 (1951)
[21] Lafont, Y., Reus, B., Streicher, T.: Continuation Semantics or Expressing Implication by Negation, Technical Report 93-21, University of Munich (1993)
[22] Laurent, O.: Intuitionistic Dual Intuitionistics Nets (2008) (submitted)
[23] Laurent, O., Regnier, L.: About translations of Classical Logic into Polarized Linear Logic. In: Proc. LICS 2003. ACM Press, New York (2003)
[24] Murthy, C.R.: Extracting Constructive Content from Classical Proofs. PhD thesis. Cornell University (1990)
[25] Okada, M.: Some remarks on linear logic. In: van Atten, M., Boldini, P., Bourdeau, M., Heinzmann, G. (eds.) One Hundred Years of Intuitionism. Birkhauser, Basel (2007)
[26] Parigot, M.: $\lambda\mu$-calculus: An Algorithmic Interpretation of Classical Natural Deduction. In: Voronkov, A. (ed.) LPAR 1992. LNCS, vol. 624, pp. 190–201. Springer, Heidelberg (1992)
[27] Plotkin, G.D.: Call-by-name, Call-by-value and the λ-*calculus*. Theoretical Computer Science 1, 125–159 (1975)
[28] Selinger, P.: Control Categories and Duality: on the Categorical Semantics of the lambda-mu calculus. Mathematical Structures in Computer Science 11, 207–270 (2001)
[29] Urban, C., Bierman, G.M.: Strong Normalisation of Cut-Elimination in Classical Logic. Fundamenta Informaticae 20, 1–33 (2001)
[30] Urban, C., Ratiu, D.: Classical logic is better than intuitionistic Logic: a conjecture about double negation translations (2008) (submitted)
[31] Zeilberger, N.: On the Unity of Duality. Ann. Pure Appl. Logic 153, 66–96 (2008)

Ontologies and Coherence Spaces

V.M. Abrusci[1], M. Romano[1], and C. Fouqueré[2]

[1] Research group "Logica e geometria della cognizione", Università Roma Tre
{abrusci,mromano}@uniroma3.it
[2] LIPN, Université Paris-Nord & CNRS, UMR7030
cf@lipn.univ-paris13.fr

1 Introduction

Semantic Web is the initiative supported by the W3C that aims to make the WorldWideWeb a place of interaction among machines – or at least among their "representatives" known as *autonomous agents* [3] – thanks to the exchange of "labelled" data. It is clearly something more complex and more interesting than today's Web, which allows only for the exchange of files and "raw" data that machines simply display on a monitor.

Within the Semantic Web initiative it is assumed that expressive formal description of data sources will lead to their interconnection throughout the WorldWideWeb *via* logical inter-definition of concepts appearing in different descriptions. According to this core idea, the research area of Knowledge Representation (KR) has been co-opted; the merge of the more promising KR models (namely ontologies) with standard markup languages to be used in the Web has been followed by the adoption of inferential engines to be plugged into knowledge bases (KB) in order to exploit the expressivity of languages designed to comply with some Description Logic (DL), namely OWL. Thus, the interconnection of datasources depends on their specification in a formal way suitable for automatic reasoning. In the meantime, the social evolution of the Web (Web2.0) has shown the effects of collective intelligence and its potential to manage large amounts of information with user-friendly tools requiring no specific (or none at all) skill in KR. However, while the top-down approach of the Semantic Web initiative looks like being too close to Artificial Intelligence, the bottom-up practices of Web2.0 lack any theoretical set-up to be really useful for autonomous agents' operations. We think that the community that is developing the technological layer of the Semantic Web should strictly cooperate with, and take advantage of, the Web2.0 communities that spontaneously provide the Web with collections of resources that are categorized (though roughly) according to some collectively developed knowledge framework.

After a brief discussion of the logical assumptions claimed or implied with KR involvement – and of their aptness to account for low-level, untrained and spontaneous categorization of resources – we propose an alternative logical framework, that of Linear Logic, which, we argue, can trigger the dynamic exchange of resources and data through different datasources. In particular we expect that it can provide the tools to describe and realize the passage from a datasource to

Lecomte and S. Trončon (Eds.): PRELUDE 2010, LNAI 6505, pp. 205–219, 2011.

any other without the need for a given-in-advance and usually "hand-made" formal description of any single datasource involved and of the mapping between every pair of them. This way, we could also get Semantic Web closer to Social Web using directly the knowledge put into Web2.0 environments (e.g. tagging spaces) without the step of ontology extraction that requires the definition of a conceptual hierarchy to be validated somehow.

2 KR and Web Resources

About ten years after the adoption of KR to support the building of the Semantic Web it is apparent that there are some major problems that make it a long and hard way to walk. To have knowledge representations suitable for use with current technologies and according to the theoretical background of Knowledge Engineering (KE), we need both given in advance formal descriptions of the datasources, and the provision of rules for the mapping between datasources to have them really exchanging data and resources. As regards the need for formal descriptions, it requires both specialists to define the domain ontologies [8] and professionals to give them the formal dressing. As regards the mapping issue, it prevents to compare any two datasources (i.e. to use them in the sense of Semantic Web) unless the mapping instructions have been provided. Put in this way, the Semantic Web looks like too ambitious as a research initiative, since it expects enormous efforts to properly define suitable ontologies, while as a commercial initiative (as it is at least partially) it follows a completely top-down and anti-dynamic approach since ontology definition invariably freezes the knowledge about something according to its understanding on the part of a restricted group of experts, thus behaving in an opposite way to the dynamic and unpredictable character of Web communities.

By contrary, within the Web2.0 movement have emerged some systems for quick and easy classification of resources, on the part of both the authors and the users, that we can generally indicate with the term *tagging*, which is studied as the form of emerging categorization called *folksonomy* [9]. In spite of ontologies, such bottom-up and popular systems are expressively very poor and usually generate no taxonomy but flat spaces where, even if there may be some hierarchy among the concepts that are identified by tag-terms, it is neither shown nor recorded when tagging.

We will not hold that tagging spaces are first class knowledge representations since they completely lack any interest for the global consistency and awareness of the complete resulting terminology or concept-scheme. Nevertheless tagging is the form of "naive categorization", made by sticking labels to resources in the Web, where folksonomy comes from. Folksonomies are aggregations of tag-terms that can be used to find out the resources that match some tag and, with it, also the corresponding concept expressed by the term, precisely as it should be for Semantic Web ontologies. Secondly, flat tag spaces are categorizing greater and greater amounts of web resources. So they seem able to do for Semantic Web that part of the work that ontologies cannot manage, i.e. large scale categorization.

If we then consider that KR has entered the Web in order to help in making machine-readable the information in it, and not primarily to develop the finest descriptions of particular worlds, we should exploit any contribution able to cooperate in achieving Semantic Web.

2.1 A Logical Model to Rely on

Being a major result of KE, ontologies and especially DL ontologies have a precise logical meaning, i.e. a definite semantic interpretation [1]. It relies on Set theory and associates to the ideas expressed by concept names (and their logical definitions) the corresponding set of "all the objects that ..." within the specific domain of interest that the ontology is designed to describe according to a usually not well defined interpretation function. The inferences that a reasoner can draw from such KBs are generated according to the axioms offered as concept definitions and possibly also presented in a "facts box" in form of assertions about individuals. Generally an OWL ontology is exposed in the WorldWideWeb in the form of an XML document which includes:

- a preamble that is the presentation of the ontology itself: i.e. its title, its identifier and possibly the identifiers of other ontologies that it is linked to. As an example we show below an extract from the OWL specification of the FOAF ontology (Friend Of A Friend) [4] – i.e. an ontology about people, their personal information and other people that they know – whose preamble looks like this:

```
<owl:Ontology rdf:about="http://xmlns.com/foaf/0.1/"
    dc:title="Friend of a Friend (FOAF) vocabulary"
    dc:description="The Friend of a Friend (FOAF) RDF vocabulary,
      described using W3C RDF Schema and the Web Ontology Language.">
</owl:Ontology>
```

- The list of concepts and their relations that constitute the conceptual domain description, structured in the form of a hierarchy of terms – this part is usually called T-box (box of the Terminology). The following extract from the same ontology shows an example about the concept Person that is subclass of the concept Agent:

```
<rdfs:Class rdf:about="http://xmlns.com/foaf/0.1/Person"
            rdfs:label="Person" rdfs:comment="A person.">
  <rdf:type rdf:resource="http://www.w3.org/2002/07/owl#Class"/>
  <rdfs:subClassOf>
    <owl:Class rdf:about="http://xmlns.com/foaf/0.1/Agent"/>
  </rdfs:subClassOf>
</rdfs:Class>
```

- and, possibly, a "store" of assertions that state the membership of some individuals into some concepts or their participation in some relationship – this part is usually called A-box (box of the Assertions), or facts box. The following is an example of A-box that provides the FOAF ontology with a very small collection of data just describing one of the authors of this paper:

```
<foaf:Person rdf:ID="me">
    <foaf:name>V. Michele Abrusci</foaf:name>
    <foaf:workInfoHomepage rdf:resource=
    "http://host.uniroma3.it/dipartimenti/filosofia/Docenti/infoprof/Abrusci.html"/>
<foaf:mbox rdf:resource="mailto:abrusci@uniroma3.it"/>
    <foaf:knows>
        <foaf:Person>
            <foaf:name>Marco Romano</foaf:name>
            <foaf:mbox rdf:resource="mailto:romano@uniroma3.it"/>
        </foaf:Person>
    </foaf:knows>
</foaf:Person>
```

The A-box is only one manner among others to ground the conceptual schema of an ontology to a collection of data – this is why one can find quite often ontologies with no facts box – and in general the presence of an A-box does not provide *per se* a model of the ontology. Thus legal inferences derived by the reasoners are those which holds in all the possible models based on the particular concrete domain (the collection of data) and interpretation function chosen. By the way we remark that usually the domain of interpretation is never given any concreteness, so that operations between ontologies do not take into account the extension of concepts but affect primarily the labels used to denote them, while the semantic interpretation is produced simply as a by-product of formalization, since it is always possible to declare which set should be the result of some operation. This would not be a problem if ontologies were not to assist interchange of data in a Web of concrete resources. But Semantic Web ontologies are precisely to describe what is within different datasources and to enable the jump from each other so that what exactly is such a domain is not a minor problem. In order to achieve a working Semantic Web, we think we should talk about *applied ontologies*, i.e. ontologies together with the collection of data or resources that are accounted for within the ontologies, so that also the operations between ontologies should be considered primarily as affecting the resources.

Now let's observe the semantic interpretation of folksonomies. To be honest, they have no definite semantic interpretation, but it seems quite easy to adopt once again Set theory and consider them as very poor ontologies without concept definitions. It would be straightforward to take some domain Δ and interpret all the resources that share a specific tag as a particular subset of Δ identified with the concept lying behind the tag-term. As regards the calculus they support, no inference is allowed, but mere resource retrieval by using tag-terms as search keys. Operations between folksonomies have not even been conceived.

2.2 One Step Further with More Structure

We propose a theoretical framework for KR specially conceived for use within the Semantic Web scenario. Behind this proposal there is a doubt about the adequacy of the current approach based on Set theory (for KR) and linguistic analysis of KBs' intensional level (for the discovery of compatible resources) w.r. to the full achievement of the Semantic Web. So, instead of focusing on conceptual schemes of ontologies, we propose to focus on the extensional level of KBs, i.e. on "real" objects, by adopting a logical framework capable to geometrically represent relations among resources, possibly discovering the concepts from resource aggregations that are actually in the Web. In particular we suggest to consider another kind of semantic interpretation that relies on structures richer than sets, the *coherence spaces*, where the interpretation of a concept (or tag-term) produces graph theoretical objects along with the determination of the extensional counterpart within the collection of resources that is the domain of interpretation.

Coherence spaces [7] are webs whose points may or may not be linked to each other according to a binary symmetric reflexive relation called *coherence*. They allow for the definition of a denotational semantic, so that we can get one also for data exchange within the Web and for a definition of operations between ontologies that is primarily focused on resources. Coherence spaces come from Linear Logic (LL) and have been the first semantic interpretation of that logical system. What is more is that it is not truth-valued semantics, useful only for talking about formulas – as it always happens w.r. to Classical Logic (LK) – but precisely denotational semantics, useful for talking also about proofs. Indeed, they offer the domain of interpretation of the objects manipulated by the logical calculus, i.e. proofs.

LL [5] is a logical system developed by imposing some restrictions on the use of structural rules for the construction of deductions within Gentzen's proof calculus for first order logic (FOL) known as sequent calculus. The affected rules are Contraction and Weakening which deal with the number of times formulas may be used within the same proof. In LL they are re-defined in form of logical rules (instead of structural) so that their usage has to be marked with specific connectives, called *exponentials*. This way everything that holds in LK also holds in LL, although LL is able to better describe what is happening in a proof: the ability to mark for which formulas it is licit to have weakening and contraction means that one can control the times resources are used. We note, by the way, that this one looks like an interesting property to have at hand while working for a Semantic Web of resources. As a consequence of the control on contraction and weakening, LL deconstructs the connectives $\wedge$ and $\vee$ doubling them in the multiplicative and additive variants, since their behaviour is different according to the possible uses of the context (i.e. the other formulas) where the formulas in which they occur are interacting with. To put it in a nutshell, the multiplicative connectives operate on the coherence space resulting as the product of the coherence spaces corresponding to the proofs of the connected formulas, while the additives on their disjoint union.

Moreover, LL has developed a geometrical representation of proofs by means of graphs called proof-nets. It exploits graph structures to compose partial proofs and provides graph properties to determine when a proof structure is correct. Such graph structures have a model in coherence spaces, where the denotation of the proof of a formula is a set of pairwise coherent points, called *clique*. The operations between coherence spaces interpret the composition of formulas and their proofs according to LL connectives. In order to account for the interpretation of ontologies through coherence spaces we need a particular class of them, with a typical support set (see below). In addition, because of the context of Semantic Web and its need for datasource integration, we propose to call the structuring relation of such coherence spaces compatibility rather than coherence. As a consequence, we propose to call this special class of coherence spaces "Ontological Compatibility Spaces" (OCS).

3 Ontological Compatibility Spaces

Let A be a semantic web ontology in a language like OWL. Let $\mathcal{L}(A)$ be the set of all the symbols for individuals, concepts and roles of A (i.e. individual symbols, unary predicate symbols, binary predicate symbols), M a set of data, ϕ a valuation through suitable mapping from $\mathcal{L}(A)$ to M, i.e. if c is an individual symbol of $\mathcal{L}(A)$ then $\phi(c)$ is a singleton set included in M, if P is a unary predicate symbol of $\mathcal{L}(A)$ then $\phi(P) \subseteq M$, if P is a binary predicate symbol of $\mathcal{L}(A)$ then $\phi(P) \subseteq M \times M$.

This defines an applied ontology $\langle A, M, \phi\rangle$ and we can represent every concept, role and individual of it by means of a special kind of coherence spaces.

Definition 1 (Ontological Compatibility Spaces, OCSs). *Based on an applied ontology $\langle A, M, \phi\rangle$, an OCS $[A, M, \phi]$ is defined as follows:*

- *its support, as usual noted $|[A, M, \phi]|$, is $M \cup (M \times M)$*
- *the coherence relation between every two points of $[A, M, \phi]$, noted $\frown_{[A,M,\phi]}$,*[1] *is assigned according to the following:*

$$\forall (x, y) \in [A, M, \phi] \times [A, M, \phi],\ x \frown_{[A,M,\phi]} y \Leftrightarrow \exists P \in \mathcal{L}(A) s.t. \{x, y\} \subseteq \phi(P)$$

As for general coherence spaces, a group a of pairwise compatible points of $[A, M, \phi]$ is called a clique, and is noted $a \sqsubset [A, M, \phi]$. More formally:

$$a \sqsubset [A, M, \phi] \Leftrightarrow |a| \subset |[A, M, \phi]| \wedge \forall (x, y) \in a \times a,\ x \frown y$$

Finally, the class of OCSs is the class of all the coherence spaces $[A, M, \phi]$ as above that represent some applied ontology, and is closed under operations on coherence spaces.

The coherence relation formalizes the notion of compatibility emerging whenever an ontology is applied to a set of data. Indeed, from an abstract point of view, the retrieved values for any predicate symbol P of A form some subset of $M \cup (M \times M)$ whose elements share with each other something more than all the other elements of $M \cup (M \times M)$. Such a property, instead of being named according to any specific symbol P occurring in the ontology, may be rewarded as the compatibility between all the points of that subset of $|[A, M, \phi]|$.

Defined as a coherence relation, the compatibility relation is reflexive, symmetric and non-transitive. Thanks to reflexivity, every point is a clique and may be considered as a minimal class of compatibility. Symmetry expresses the core

[1] For sake of clarity we may note simply $\frown$ when the coherence space is obvious.

of the idea of compatibility, since for compatibility we mean the possibility to put two "objects" together based on their sharing of some common property – not necessarily an expressed one – and such a commonality has to be inevitably a reciprocal fact. Non-transitivity prevents the overwhelming distribution of compatibility that would mix up different Ps of A whenever they have some common points. However it does not forbid to make a clique of points that inherit their compatibility due to the hierarchy in which Ps are organized.

As with coherence spaces, an OCS $[A, M, \phi]$ can be represented through a graph $G(V, E)$: $V = |[A, M, \phi]|$ is the set of nodes and $E \subseteq |[A, M, \phi]| \times |[A, M, \phi]|$ is that of edges with the constraint that for every two points x, y of $|[A, M, \phi]|$ we have $\{x, y\} \in E \Leftrightarrow x \frown y$ that is to say $\{x, y\} \in E \Leftrightarrow \exists P \in \mathcal{L}(A)$ s.t. $\{x, y\} \subseteq \phi(P)$. Based on the definition of coherence, when looking at the graph of an OCS, a clique $a \sqsubset [A, M, \phi]$ turns out to be a completely connected subgraph.

Remark. We observe that the representation of ontologies as OCSs successes:

1. for every symbol $s \in \mathcal{L}(A)$ (that is every individual or concept or role of $\langle A, M, \phi \rangle$) $\phi(s) \sqsubset [A, M, \phi]$.
2. following the inverse direction, every clique of the OCS is the denotation of:
   - some concept or role or individual of $\langle A, M, \phi \rangle$;
   - or a new concept not identified by any predicate of $\mathcal{L}(A)$ which corresponds to either a subconcept of some other concept specified in $\langle A, M, \phi \rangle$ or to a new concept "spreading" over different concepts already identified in $\mathcal{L}(A)$.[2]

As every coherence space OCSs satisfy also the properties of down-closure and binary completeness, i.e.:

- if $a \sqsubset [A, M, \phi]$ and $a' \subset a$ then $a' \sqsubset [A, M, \phi]$
- if $B \subset [A, M, \phi]$ and $\forall a_1, a_2 \sqsubset B$ $(a_1 \cup a_2 \sqsubset B)$ then $\bigcup B \sqsubset [A, M, \phi]$

Talking about ontologies this means that: i) any subconcept of a concept has to be considered as a concept, even if there is no name for it in $\mathcal{L}(A)$, since we care only about cliques of compatible points; and ii) when the union of two concepts still forms a clique for every two concepts in (a subset of the concepts of) an OCS, all those concepts together form a concept too – i.e. a whole (branch of an) ontology forms a concept when also the union of any two of its concepts is still a concept.

Example. Based on the same ontology of which we have shown above some extracts, we now show what looks like its representation as an OCS. We remark that what is going to be shown in an OCS are the actual objects accounted for in the ontology, grouped as cliques as long as one concept fits for all the objects of a clique.

---

[2] This is a possible way to discover new concepts that is especially interesting with folksonomies.

Regarding our very simple ontology (cf. section 2.1) we have the following "points of information": two Persons, their Email accounts and the Web-page of one of them. So we have an OCS that shows (step by step):

- a clique of Persons

It counts two points because in the few lines of the example A-box above we have just two resources bearing the label `foaf:Person`

- and a clique of Email accounts

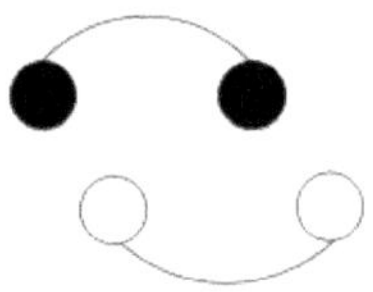

Again two points as in the "code" above we have two strings introduced by the label `foaf:mbox`

- and a clique of Web-pages

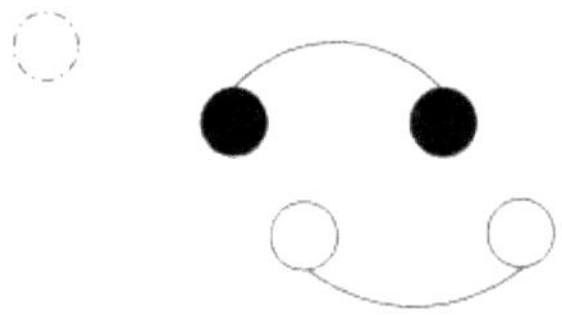

It counts just a single point since in the example A-box there is only one string with the label `foaf:workInfoHomepage`

- and also a clique of pairs made of a Person and his Email account

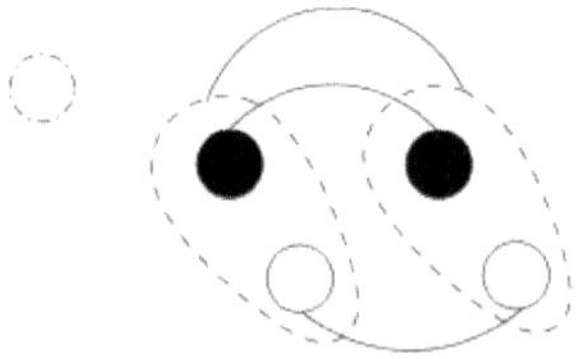

This clique counts two points (dashed circles in the figure) because in the A-box we have two pairs made of a resource labelled `foaf:Person` and a string attributed to it by the role `foaf:mbox`

As the example suggests, concepts look just like completely connected subgraphs while the roles of the ontology let one take pairs of points instead of single points.

4 Operations between Ontologies

The best that the interpretation of ontologies as OCSs may bring to Semantic Web is the consequent interpretation of operations between ontologies as operations between coherence spaces, which may provide a theoretical frame to describe the use of ontologies as sources of information for Web-agents. Usual operations with ontologies range from union of ontologies to module extraction (that may be considered as the inverse of the union), ontology mapping and ontology merging (which is the result of the combining of mapping and union), while coherence spaces interpret all the operations available within the logical system of LL, so that we have some "new" operations with ontologies that is worth to further study as to define their possible role in describing actual interactions (information exchanges) between agents in the Web. For the time being, let's have a glance on some of these operations.

– The dual of an ontology

The dual of $[A, M, \phi]$ is a coherence space noted $[A, M, \phi]^{\perp}$ with the same web, where the compatibility has been replaced by incompatibility ($\asymp$), so that all compatibilities holding in $[A, M, \phi]$ are removed and new compatibilities are assigned in $[A, M, \phi]^{\perp}$ only to pairs of points which are either identical or not together a clique of $[A, M, \phi]$. The cliques of $[A, M, \phi]^{\perp}$ may be found according to the correspondence between a coherence space and its dual: $a \sqsubset [A, M, \phi]^{\perp} \Leftrightarrow \forall (x, y) \in a \times a \; x \asymp_{[A,M,\phi]} y$ so that a clique of $[A, M, \phi]$ and one of $[A, M, \phi]^{\perp}$ can share at most one common point. Duality of coherence spaces provides us with an unusual idea of negation for ontologies, one that does not deal with an operator for concept definition (complementarity) but one that affects the whole ontology: compatibility becomes incompatibility and the other way around, a phenomenon that can be understood as a radical change of the observational standpoint about the knowledge described in the ontology.

- The union of ontologies ($\oplus$)

The sum $\oplus$ of two OCSs is the disjoint union of their webs; nothing is added to the compatibility. Thus, each point of the web of $[A, M, \phi]$ remains incompatible with any point of $[A', M', \phi']$. We observe that such a sum corresponds to a special case of the usual operation of (controlled) ontologies union, where concepts from ontology A' are to be added to concepts from ontology A provided that no mixing of concepts may occur. In fact, from the graph of $[A, M, \phi] \oplus [A', M', \phi']$ one can recover only cliques which are either cliques of $[A, M, \phi]$ or of $[A', M', \phi']$.

- The dual of the sum (&)

Since duality is characteristic of coherence spaces we consider also the dual of the operations between them. The dual of $[A, M, \phi] \oplus [A', M', \phi']$ is $[A, M, \phi]^{\perp} \& [A', M', \phi']^{\perp}$, whose cliques are all the cliques of $[A, M, \phi]^{\perp}$ and of $[A', M', \phi']^{\perp}$ along with the union of all the cliques of $[A, M, \phi]^{\perp}$ with the cliques of $[A', M', \phi']^{\perp}$. The dual of the union of two ontologies is far less comfortable to deal with since, as it is for the same notion of dual, it is quite unknown to ontology practice. Nevertheless, while the union of ontologies simply adds the list of concepts in A' to concepts in A, we remark that with the dual of the union we consider a new unified space where the points of $[A', M', \phi']^{\perp}$ are always compatible with the points of $[A, M, \phi]^{\perp}$.

- Mapping ontologies in ontologies ($\multimap$)

The expression "ontology mapping" covers a variety of quite different techinques to compose two (or more) ontologies in a single ontology where possibly new relations between concepts from the two ontologies are shown. Usually it is operated only caring for the signature (i.e. the concept names and their specifications) of the involved ontologies. We want to deal with applied ontologies so we should describe what happens with the collections of tagged data. This is to say we consider the mapping of a collection of data as a whole, concerning not only its formal specification and presentation.

Mapping an ontology in another one actually produces a third ontology (which sometimes is called the "reconciled ontology") then we can represent this as follows: two applied ontologies $\langle A, M, \phi\rangle$ and $\langle A', M', \phi'\rangle$ are mapped into a third ontology $\langle A'', M'', \phi''\rangle$ which is actually produced taking a copy of every point of $[A, M, \phi]$ and $[A', M', \phi']$ into the new space $[A'', M'', \phi'']$ and preserving all the compatibilities of each point with the others in the OCS it comes from. Being a point-by-point projection in the new OCS, such a process can easily be seen as a linear function $F : [A, M, \phi] \oplus [A', M', \phi'] \multimap [A'', M'', \phi'']$.

What is really going to be the resulting OCS is by far the most interesting part, even though it is also the less predictable. Having no *a priori* information about what is in $[A, M, \phi]$ and $[A', M', \phi']$ we should assume that the resulting space is either $[A, M, \phi] \oplus [A', M', \phi']$ or $[A, M, \phi] \,\&\, [A', M', \phi']$. For the general case we have that $[A, M, \phi] \oplus [A', M', \phi'] \subseteq [A'', M'', \phi''] \subseteq [A, M, \phi] \,\&\, [A', M', \phi']$, what gives an idea of the increase of mutual compatibility between the two spaces. Indeed, the actual appearence of $[A'', M'', \phi'']$ may be much more complex than the cases of $\oplus$ and & due to the mixing of some cliques of $[A, M, \phi]$ with some others of $[A', M', \phi']$ that is possible thanks to the discovery of new inclusion relations between concepts of the two original ontologies. This is exactly the kind of result that is expected with the adoption of empirical techniques – that typically rely on linguistic analysis and implement some form of machine learning – for what is usually called the "discovery of mappings" between ontologies. Such an empirical result is beyond the action of the mapping function and is what we suggest to adopt Ludics [6] for during the further development of our research. New maximal cliques that possibly can be found within $[A'', M'', \phi'']$ – i.e. those resulting from the mixing of the projections of cliques of $[A, M, \phi]$ and $[A', M', \phi']$ – will be the new concepts proper to the new OCS, i.e. what is eventually the increase of knowledge produced by the mapping operation, which is a mapping function together with some empirical acquisition.

As regards the details of our mapping function F, we define it as follows (noting for short $[A, M, \phi] \oplus [A', M', \phi']$ as X and $[A'', M'', \phi'']$ as Y): i) $F(\emptyset) = \emptyset$; ii) $\forall x \in |X|\ \exists y \in |Y| s.t. F(x) = y$; iii) $x \neq y \rightarrow F(x) \neq F(y)$; iv) $\forall a \sqsubset X\ F(a) = \bigcup[F(x)|x \in a]$. Such a function is linear: given any two concepts of an applied ontology $\langle A, M, \phi\rangle$, e.g. $a, b \sqsubset [A, M, \phi]$, we have that $F(a) \cup F(b) = F(a \cup b)$. This is straightforward if one considers that F pointly reconstructs in the new OCS exactly what is in each of the original ones. Also stability can be easily proved, and together with the preservation of inclusions and unions, we have that F safely preserves the relations between concepts.

Having an interesting tool to account for the mapping of collections of data, the open challenge will be to develop some technique suitable for the account of the empirical findings about concepts – i.e. the additional mixing of cliques in a mapping of ontologies that is due only to their actual "meaning", or value.

Example. Let's have a look to what is the representation of such operations in form of OCSs. We tackle here just the two cases that are familiar to usual work with ontologies. The applied ontologies we will use in the following are the A-box we have already shown above (in section 2.1) and the one that we show below, which is based on another ontology for personal information management (PIM) [2], developed by a small group of researchers involved in the W3C's Semantic Web Initiative[3].

As regards our second A-box, you may note that its language is not OWL, but RDF (Resource Description Framework). In fact, since it is just an A-box and it makes no use of OWL special terms, it can be expressed in a more basic language for data description (which by the way is also more largely supported over the Web). The axioms and concept definitions that form the T-box – the conceptual description – for which this A-box provides a collection of data, are not listed here, but are recalled by the import mechanism of namespaces. Indeed an ontology is not to be copied in its entirety every time one has to refer to one or more concepts that the ontology defines; it needs just to "import" that ontology, that is to put a link to it. Once an ontology is imported one can freely use everything that is stated in it thanks to the technique of namespaces and the use of a special prefix (in bold in our examples) to refer to before each term in the ontology. This allows to unambiguously refer to each term's specification wherever may be the file containing such specification. Links to other ontologies are stated in the preamble, where a special prefix is associated to each namespace that will be recalled in the ontology.

```
<rdf:RDF
  xmlns:rdf="http://www.w3.org/1999/02/22-rdf-syntax-ns#"
  xmlns:rdfs="http://www.w3.org/2000/01/rdf-schema#"
  xmlns:pim="http://www.w3.org/2000/10/swap/pim/contact#">
   <pim:Person rdf:ID="me">
      <pim:name>Marco Romano</pim:name>
      <pim:title>Mr</pim:title>
      <pim:mailbox rdf:resource="mailto:mromano@uniroma3.it"/>
      <pim:homePageAddress rdf:resource="http://www-lipn.univ-paris13.fr/~romano/"/>
   </pim:Person>
</rdf:RDF>
```

First of all we consider the simplest (trivial) way to put together the information within two different (applied) ontologies, i.e. what we have called their sum ($\oplus$).

As the figure shows, facts asserted in the two ontologies are just put side by side, no mixing of information can take place with such operation. The result is simply to have all the cliques from the two OCSs collected in a new single OCS.

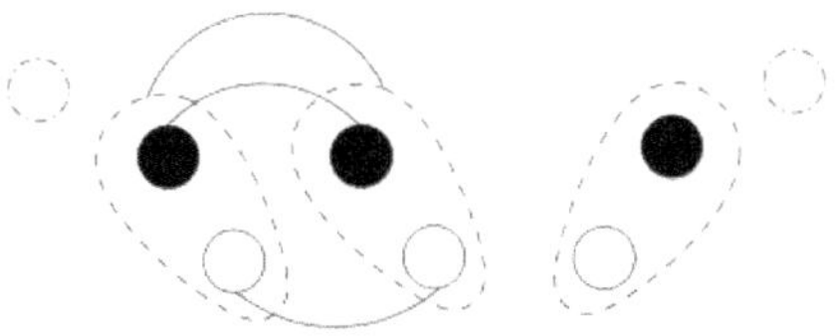

[3] The ontology is available on-line at http://www.w3.org/2000/10/swap/pim/contact. Among the people who have been working on it since early 2000s let us mention Tim Berners-Lee, co-inventor of the WorldWideWeb.

Then we consider the more interesting operation of mapping. We recall that this operation aims to establish equivalence between labels specified in different ontologies whenever such labels are intended to describe the same concept. How to discover when actually two labels may be used for the same concept is the crucial point, but for the present example we will rely on the very simple statement of mapping between the two ontologies that is provided within the definition of the concept Person in the FOAF ontology. Indeed it states that `foaf:Person` is a subclass of `pim:Person` and this allows to infer that everything with the label `foaf:Person` can be used as if it had the label `pim:Person`. Now we show how the mapping established as a T-box axiom affects the actual information available in the applied ontologies.

The mixing of cliques affects only the cliques corresponding to the labels `foaf:Person` and `pim:Person` even though, reasonably, if one considers the actual meaning of the other facts described in the two A-boxes, also other cliques should mix (e.g. those corresponding to `foaf:mbox` and `pim:mailbox`).

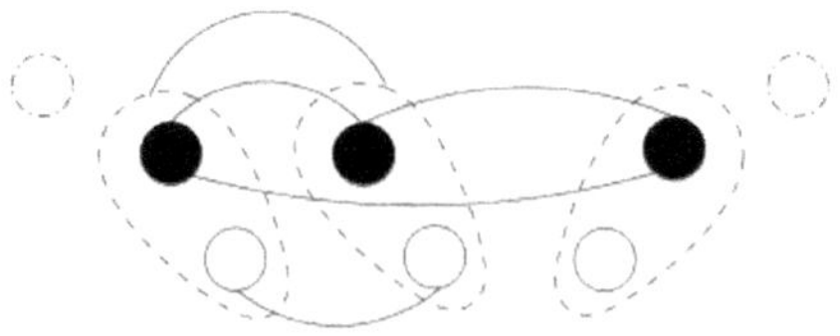

To say all the story, one could expect that there is also some form of collapse of the graph in the mapping because both the ontologies state something about an instance of the concept Person that is named `Marco Romano` – and the same holds for his email address. But this is quite far to achieve with nowadays techniques although it may look so trivial to a human being to recognize that this is indeed the same person. Since mappings are established only for T-box elements and then, if necessary, projected over A-boxes – and since here we are given only the mapping instruction for the two labels about persons – the other cliques simply cannot mix, and there is no way for the single points to be identified as referring to the same "real" objects. Obviously, what we look for in the following of our research is a way to have cliques conveniently mixing not only according to what is stated as special mapping rules (or occasional inter-definitions of concepts), but also depending on other factors like e.g. the use of the actual resources on the part of some user.

5 OCS vs. Ontology

How may OCSs be of better use than ontologies w.r. to the social (and pragmatic) aspects of Web2.0? In fact we have designed OCSs specially for representing ontologies, but actually it looks like folksonomies can benefit even more than ontologies from OCSs.

Nowadays when looking at the set of tag-terms adopted within a community it is expected to reconstruct a formal ontology out of that, establishing a neat and formal hierarchy among concepts, useful for resource retrieval according to the traditional top-down approach of progressive specification. The major side effect of such a reduction is the loss of the dynamic aspect of Web2.0.

We may observe that building the Semantic Web by means of ontologies requires predefined sets of metadata (the schemes) to be adopted and respectfully obeyed. Their usefulness – and the wealth of Semantic Web itself as the workplace of autonomous agents – will depend in fact on the number of resources whose set of metadata matches one or more of the predefined schemes so that programs specially written according to the same scheme(s) will be able to use those resources. So the Web of data that W3C indicates turns out to be something like a giant database where a neat definition of the logical scheme (even composed of many different ontologies) can be achieved only thanks to the standardization of the metadata tags to be used to describe resources.

In the opposite direction goes the practice of free tagging, so that folksonomies emerge as everlasting works in progress where concept-terms institution and resource description and classification always happen in the same time, with no hope for standardization. In fact, when people tag they freely choose and establish their own categories in an unending process of ontology elaboration. Moreover, while using their very personal categories people also express their own "world's understanding" so that tagging spaces are not only useful for classification, but also convenient for collective intelligence to share knowledge.

We remark that tagging spaces publish enormous amounts of resources with some kind of classification while providing a cognitive framework that has not the claims of ontology but it is powerful enough to let one recognize and find classes of resources that are compatible, i.e. similar to some extent. Maybe such a cognitive framework is a lower quality contribution, w.r. to formal DL ontologies, to have the content of the Web surely recognizable, but it seems to show a more feasible way to do that. In fact, since it relies on no pre-emptive requirements, no standardized label to tag resources, it preserves the dynamical behaviour of Web2.0 and lets Semantic Web to be a "common people affair" as it has been for last years for WorldWideWeb, and for its big boom.

Instead of the usual techniques for tags clustering and concept extraction, we may exploit OCSs in order to recover from tagging spaces a description of the resources in a datasource that is formal enough to be useful for data exchange but that does not need *ad hoc* specification of a conceptual hierarchy. We look only for compatibility between resources observing the connections given within an OCS. Our proposal is to give more logical dignity to flat tagging spaces without rising too high in formal complexity so as to prevent large contribution from common web-users. We simply aim to describe flat spaces in such a way that it makes sense to talk about operations between them. In order to have an OCS out of a flat tagging space we need nothing more than what has been stated for ontologies, since we consider it as a very simple ontology, with just some niceties: the language $\mathcal{L}(F)$ of such a folksonomy F will be the pair (T, I) where T is the set of tag-terms and I the set of the resource identifiers (the WorldWideWeb standard Universal Resource Identifiers). The web of the OCS $[F, I, \phi]$ will be $|[F, I, \phi]| = I$. ϕ is the valuation of a tag-term $\mathbf{t}$ on the set $\phi(\mathbf{t}) \subseteq I$ of the resources tagged with $\mathbf{t}$ – i.e. it is a query. The compatibility relation is slightly revised as $x \supset_{[F,I,\phi]} y \Leftrightarrow \exists \mathbf{t} \in \mathcal{L}(F)$ s.t. $\{x, y\} \subseteq \phi(\mathbf{t})$.

6 Conclusions

In conclusion, we recall some of the most challenging aspects that are to be further investigated in our research along with the results that we can see at present. Recurring to coherence spaces to describe operations between ontologies (and folksonomies), we have to face some deep questions that emerge from aspects inner to LL and coherence spaces theory. Indeed, coherence spaces and their cliques represent proofs of (multisets of) formulas (from here on we use formula while meaning multiset). Each coherence space offers the place for the denotation of one formula and shows a correspondence regarding a clique, a formula and its proof, while operations between coherence spaces provide the denotation for the calculus when composing formulas. When adopting coherence spaces for ontologies, we have to find the right place to put other elements in this correspondence, i.e. concepts and resources, together with ontology. We have to clearly set the circle of correspondences and a first attempt is to compare the formulas appearing in the right-hand side of a sequent in LL with an ontology seen as the union of its concepts, so that the proof satisfying one of the concepts satisfies also the whole ontology.

Another point is to find out precisely what may be the meaning – and usefulness for Semantic Web – of certain operations that OCSs make possible for ontologies (e.g. the dual) that at present have no apparent meaning w.r. to standard use of ontologies.

For the time being, we propose something like an alternative description language to represent datasources. With such a language we are given the easy of use of tagging space, the ontology/folksonomy extraction procedure described above and a calculus based on LL connectives, that are a little finer than those usually adopted working with ontologies. We have already mentioned the doubling of some connectives ($\wedge$ and $\vee$, in their multiplicative and additive formulation) in LL, together with the appearing of two new connectives, the exponentials, that control the use of contraction and weakening, i.e. the use of resources. Without doubt it is worth to assess the usefulness of such "logical controllers" in order to account for operations in the Web that involve resources.

Finally, as the most interesting perspective deriving from the present research, we indicate the involvement of Ludics as the theory able to account for the discovery of compatibility between resources (points of OCSs) in the Web, even (and most of all) between resources coming from different datasources and counted in different OCSs.

References

[1] Baader, F., Calvanese, D., McGuinness, D., Nardi, D., Patel-Schneider, P. (eds.): The Description Logics Handbook: Theory, Implementation and Applications. Cambridge University Press, Cambridge (2003)

[2] Berners-Lee, T., Connolly, D., et al: PIM ontology, `http://www.w3.org/2000/10/swap/pim/contact`

[3] Berners-Lee, T., Hendler, J., Lassila, O.: The Semantic Web, Scientific American (May 2001)

[4] Brickley, D., Miller, L.: The Friend Of A Friend (FOAF) Project, `http://www.foaf-project.org/`

[5] Girard, J.-Y.: Linear logic. Theoretical Computer Science 50, 1–102 (1987)

[6] Girard, J.-Y.: Locus solum: from the rules of logic to the logic of rules. Mathematical Structures in Computer Science, 301–506 (2001)

[7] Girard, J.-Y., Lafont, Y., Taylor, P.: Proofs and Types. Cambridge University Press, Cambridge (1989)

[8] Gruber, T.: Towards principles for the design of ontologies used for knowledge sharing. Int'l Journal of Human and Computer Studies 43(5-6), 907–928 (1995)

[9] Vander-Wal, T.: Folksonomy (2007), `http://vanderwal.net/folksonomy.html`

Author Index

GPSR Compliance

The European Union's (EU) General Product Safety Regulation (GPSR) is a set of rules that requires consumer products to be safe and our obligations to ensure this.

If you have any concerns about our products, you can contact us on ProductSafety@springernature.com

In case Publisher is established outside the EU, the EU authorized representative is:

Springer Nature Customer Service Center GmbH
Europaplatz 3
69115 Heidelberg, Germany

Batch number: 10844619

Printed by Printforce, the Netherlands